G000047585

Gorilla

Animal
Series editor: Jonathan Burt

Already published

Ant Charlotte Sleigh · *Ape* John Sorenson · *Bear* Robert E. Bieder
Bee Claire Preston · *Camel* Robert Irwin · *Cat* Katharine M. Rogers
Chicken Annie Potts · *Cockroach* Marion Copeland · *Cow* Hannah Velten
Crow Boria Sax · *Dog* Susan McHugh · *Donkey* Jill Bough
Duck Victoria de Rijke · *Eel* Richard Schweid · *Elephant* Daniel Wylie
Falcon Helen Macdonald · *Fly* Steven Connor · *Fox* Martin Wallen
Frog Charlotte Sleigh · *Giraffe* Edgar Williams · *Gorilla* Ted Gott and Kathryn Weir
Hare Simon Carnell · *Horse* Elaine Walker · *Hyena* Mikita Brottman
Kangaroo John Simons · *Lion* Deirdre Jackson · *Lobster* Richard J. King
Moose Kevin Jackson · *Mosquito* Richard Jones · *Ostrich* Edgar Williams
Otter Daniel Allen · *Owl* Desmond Morris · *Oyster* Rebecca Stott
Parrot Paul Carter · *Peacock* Christine E. Jackson · *Penguin* Stephen Martin
Pig Brett Mizelle · *Pigeon* Barbara Allen · *Rat* Jonathan Burt
Rhinoceros Kelly Enright · *Salmon* Peter Coates · *Shark* Dean Crawford
Snail Peter Williams · *Snake* Drake Stutesman · *Sparrow* Kim Todd
Spider Katja and Sergiusz Michalski · *Swan* Peter Young · *Tiger* Susie Green
Tortoise Peter Young · *Trout* James Owen · *Vulture* Thom Van Dooren
Whale Joe Roman · *Wolf* Garry Marvin

Gorilla

Ted Gott and Kathryn Weir

REAKTION BOOKS

Published by
REAKTION BOOKS LTD
33 Great Sutton Street
London EC1V ODX, UK
www.reaktionbooks.co.uk

First published 2013
Copyright © Ted Gott and Kathryn Weir 2013

All rights reserved

No part of this publication may be reproduced, stored in a retrieval system or transmitted, in any form or by any means, electronic, mechanical, photocopying, recording or otherwise without the prior permission of the publishers.

Printed and bound in China by C&C Offset Printing Co., Ltd

British Library Cataloguing in Publication Data
Gott, Ted, 1960 July
Gorilla. – (Animal)
1. Gorilla
I. Title II. Series III. Weir, Kathryn Elizabeth, 1967–
599.8'84-DC23

ISBN 978 1 78023 029 0

Contents

1 Lascivious Beast or Shy Vegetarian?

> I do not mean to ascribe to him the highest attributes of man, or exalt him above the plane to which his faculties assign him; but there are reasons to justify the belief that he occupies a higher social and mental sphere than other animals, except the chimpanzee.
>
> Richard L. Garner, *Gorillas and Chimpanzees* (1896)[1]

In 1859 Charles Darwin's landmark publication *On the Origin of Species by Means of Natural Selection, or the Preservation of Favoured Races in the Struggle for Life* polarized scientific, religious and social debate around the world. Its revolutionary synthesis of knowledge about evolution had a profound and immediate impact upon contemporary thought, as the scientific and theological worlds argued the case for creationism versus evolution. The gorilla was central to this discussion of humanity's place in the biological order in relation to primates.

The link between primate studies and the attempt to define the human is the subject of Donna Haraway's groundbreaking monograph *Primate Visions: Gender, Race, and Nature in the World of Modern Science* (1989). Haraway explores the ways in which, in Europe, America and Japan, gorillas and other apes 'have been subjected to sustained, culturally specific interrogations of what is means to be "almost human"', and how stories about primates are simultaneously stories about 'the relations of nature and culture, animal and human, body and mind, origin and future'. The constructed nature of scientific knowledge is particularly clear in relation to gorillas and other apes, positioned as they are as humanity's means of creating its own identity and uniqueness, a phenomenon Haraway calls 'simian orientalism'.[2]

Mountain gorillas, February 2008.

As imaged in art, science, film and popular culture, the gorilla has occupied a prime position in explorations of humanity's animal nature. The social history of the gorilla reflects how quickly scientific study and debate can be drawn into popular culture. In nineteenth-century accounts of observing gorillas in the wild lie the origins of Tarzan, King Kong and many other literary and artistic representations of humanity's relationship with apes. The gorilla has provided a screen upon which to project fears of sexuality and uncontrolled drives, theories of criminality, and narratives of human and primate difference. Little wonder that knowledge of the true behaviour of the gorilla remained extremely limited outside of central Africa until the revelatory field researches of George Schaller and Dian Fossey in the 1950s and '60s.

Gorillas live in primary and mixed forests in regions of central Africa, avoiding human activity and roads, but today they often come in contact with humans in disturbed forests and when taking food from gardens near the forest edge, as well as when they fall into snare traps. Africans living in areas of gorilla habitation are very aware of competing claims on limited forest resources and land. Given local people's claims to viable subsistence and the interests vested in the animals by parties including conservationists and government tourism agencies, a complex political situation surrounds the four sub-species of gorilla in the ten gorilla-range countries where they are found. These sub-species are now generally considered to belong to two separate species: the Eastern gorilla (*Gorilla beringei*) and the Western gorilla (*G. gorilla*). Western gorillas include the Western lowland gorillas (*G. gorilla gorilla*, critically endangered and estimated at less than 200,000 in 2010, and with the most widespread habitats across

Large male silver-back gorilla sitting.

Western lowland gorilla, Cincinatti Zoo, July 2005.

Nyango, a Cross River gorilla (*Gorilla gorilla diehli*), Limbe Wildlife Centre, Limbe, Cameroon, November 2006.

sections of Angola, Cameroon, Central African Republic, Republic of Congo, Gabon, Equatorial Guinea and Nigeria), and Cross River gorillas (*G. g. diehli*, inhabiting the rainforest region across the Cameroon–Nigeria border, critically endangered with a population estimated at less than 300 in 2010). Eastern gorillas include

Eastern lowland gorillas (*G. g. graueri*, critically endangered with less than 5,000 believed to be in existence, mainly in the Democratic Republic of the Congo or DRC) and mountain gorillas (*G. g. beringei*, inhabiting a small mountainous area of the DRC, Rwanda and Uganda, critically endangered, with a population of 790 counted in a census in 2010).[3]

Gorillas, like humans, are classified in the order primates. They are believed to have evolved along paths which diverged from a common anthropoid lineage approximately seven million years ago (rather than being descended one from the other).[4] The gorilla is a primarily peaceful animal but may have great strength and size. An adult male of the species, 1.7 metres tall, can weigh more than 200 kilograms, with an arm span of 2.3 metres. The gorilla's pelage or hair is usually dark brown or black, although Western lowland gorillas may also have reddish

Mountain gorilla, September 2011.

colouring on the head. Mature males develop grey hair on their backs and flanks, leading to their being termed 'silverbacks'. White gorillas, such as those popularized in the low-budget films *White Pongo* and *The White Gorilla* (both 1945) are exceedingly rare. Snowflake, a white-haired and blue-eyed Western lowland gorilla who lived at Barcelona Zoo from 1964 to 2003, remains unique in his family – none of his 22 offspring inherited their father's pigmentation.[5] In maturity the skulls of gorillas become markedly prognathous, with males also developing a pronounced saggital crest bone which may serve to support their massive facial muscles.[6] Like other male primates apart from the human male, the male gorilla possesses a baculum, or penile bone. The gestation period for pregnant female gorillas is slightly less than for humans, while mothers usually nurse their offspring for three to four years. Gorilla sexuality is complex, and involves varying degrees of presexual play and affection.[7] In the wild gorillas enjoy polygamous sexuality, with each gorilla band or family

Gorilla impersonator Ray Corrigan in *White Pongo* (1945).

Male gorilla guarding a female gorilla.

group usually headed by a male silverback and containing a number of mature females and their offspring, including immature males. New groups may be established when females break away to join a different male silverback gorilla. Highly social, gorillas are relatively stable in these social units, which range in size and configuration; nevertheless, groups consisting solely of males or females respectively are not unknown, and numerous males live a solitary 'bachelor' existence. A silverback in charge of others maintains order within his group, sorting out 'family' disputes and warding off intrusion or attacks from outsiders.

Gorillas dwell mostly on the ground, although they are capable of climbing trees – enabled by their long arms and the opposable big toe on each foot, which permits them to grasp tree trunks and branches with ease. Although gorillas can walk bipedally for short distances, they predominantly walk on all fours upon their front knuckles. At night they build nests by flattening and intertwining plants. These are usually built upon the ground, although females and young gorillas often construct their night-time nests in the forks of trees. Gorillas are almost completely vegetarian, subsisting on a widely varied diet of more than 200 forest plants, fungi and fruits that they break down with their strong slicing molars. More than 20 kilograms of such foods per day are required to sustain an adult of the species, and they occasionally supplement their diet with ants, termites and mineral-rich soil. Communication between members of the species is conveyed through what is now recognized as a complex system of gestures, facial expressions and vocalizations. The celebrated breast-beating of the adult male is part of a defensive mechanism designed to signal his immense strength and power, thereby discouraging aggression such as that threatened, say, by a lone male gorilla in search of female partners.

There is a curious dichotomy in our knowledge of and contact with the gorilla today – the hundreds of gorillas that populate zoos worldwide are primarily lowland gorillas, while the majority of the field observations and studies of the gorilla in the wild that have been conducted since the inception of this kind of research in the 1950s have been devoted to the mountain gorilla.[8] Apart from local communities and a small number of fortunate tourists to their habitats, the wider world experiences the mountain gorilla at second hand through publications and documentary films.

Created for the fourteenth Secession exhibition in 1902, Gustav Klimt's *Beethoven Frieze* shocked Vienna with its frank depiction

Young Western lowland gorilla picking flowers.

of non-idealized nudity and unashamed sexuality. The centre of this monumental homage to Beethoven's Ninth Symphony, in which suffering humanity must pass through the realm of the Hostile Forces before finding truth in art and love, was presided over by a gigantic winged gorilla. This was Klimt's conception of Typhon, the deadliest monster from ancient Greek lore, who is accompanied in the *Beethoven Frieze* by his daughters, the three Gorgons, as well as by Lasciviousness, Wantonness and Intemperance. Shocked by this conjunction of a gorilla with nude figures flaunting their sexuality, critics labelled the work 'obscene', 'pathological' and 'painted pornography'.[9] Believed by Europeans at the time to be the most dangerous of animals, and also associated with unbridled libidinous desire, the gorilla was a powerful choice for Klimt's image of the most frightening creature in Greek mythology.

Gustav Klimt, *Beethoven Frieze*, 1902 (detail). Secession building, Vienna.

Only the previous year a German commercial traveller, Herr Paschen of Schwerin, had shot what was reported to be the world's largest gorilla. Images of this 'monster of immense proportions' were syndicated worldwide, and the toothless grin seen on Klimt's Typhon reflects the frightening rictus found on the face of this and other slain gorillas.[10] The skin and skeleton of the animal, which had been killed in Cameroon in what was then German West Africa, were sold to the Umlauff Museum in Hamburg where, taxidermied and mounted in an aggressive stance with its mouth open in a mighty snarl, it was reported to be 'a stupendous sight . . . drawing great crowds of interested spectators'.[11] Paschen himself called his kill a 'fearful monster', and the *Los Angeles Times* informed its readers that here was the very beast that had recently committed an unspeakable sadistic and sexualized atrocity upon an African woman who had been dragged away by a mighty man-like creature when she strayed from a village at dusk and then 'torn to pieces, not as a beast of prey would tear a victim, but in an indescribably horrible way, making the white men who heard the story think instantly of the stories of Jack the Ripper'. Paschen 'did not hunt the beast for the delight of the chase', the newspaper declared; 'men do not hunt the devil if they can help it, and a devil the gorilla is'.[12]

The questionable reliability of early European and North American travellers writing on African beliefs about the gorilla in no way prevented their accounts from spawning a myriad of imaginative progeny. Lurid tales like that reportedly relating to Paschen's 'monster' were embellished to feed preconceived ideas of the fearsome gorilla. Yet some also retained the kernel of a story told by central African peoples: it has been speculated that ape abduction tales may have first developed amongst Africans fearing abduction into slavery in the nineteenth century.[13] Few

Herr Paschen's gorilla, shot in 1901.

detailed or reliable accounts have been published of how the local peoples who have regularly encountered and hunted gorillas conceive of them or refer to them in oral histories and mythology. What has been recorded suggests that the animals have held a richly symbolic position in the experience and imagination of those who live in proximity to their habitats. These animals that physically so closely resemble humans have engendered tales of lust, greed and struggles for power that reflect human fears and desires.

Stories are told across central Africa of people transforming into gorillas or being reincarnated in gorilla form. In 1883 the French adventurer Louis Jacolliot wrote of how he was told by Congolese peoples that the gorilla was not like other apes, for the bodies of some of the fiercest and strongest of these creatures were animated by the spirits of dead people, who were condemned to live out a second life on earth in this new guise as punishment for crimes they had committed during their human existence. The human-gorilla hybrids were believed to be invincible, protected by a special charm that made their bodies resistant to spears and bullets. These killer gorillas were also vampiric stranglers of any solitary travellers unlucky enough to cross their paths.[14] Other men, Jacolliot wrote, cursed by sorcerers, were thought to transform into gorillas even before their deaths, shifting shape by moonlight like werewolves.[15] In his hugely popular *Explorations and Adventures in Equatorial Africa* (1861), French-American writer Paul du Chaillu recounted tales told to him in Gabon by the 'Mbondemo' (Seke) people, in fireside conversation with whom 'several spoke up and mentioned names of men now dead whose spirits were known to be dwelling in gorillas . . . in these "possessed" beasts, it would seem that the intelligence of man is united with the strength and ferocity of the beast.'[16] In 1889 an American journalist, describing a gorilla hunt with a

Paschen's gorilla in the Umlauff Museum, Hamburg, photogravure published in *The Sphere* (20 July 1901).

Cover of *Journal des voyages et aventures* (19 August 1948), woodcut.

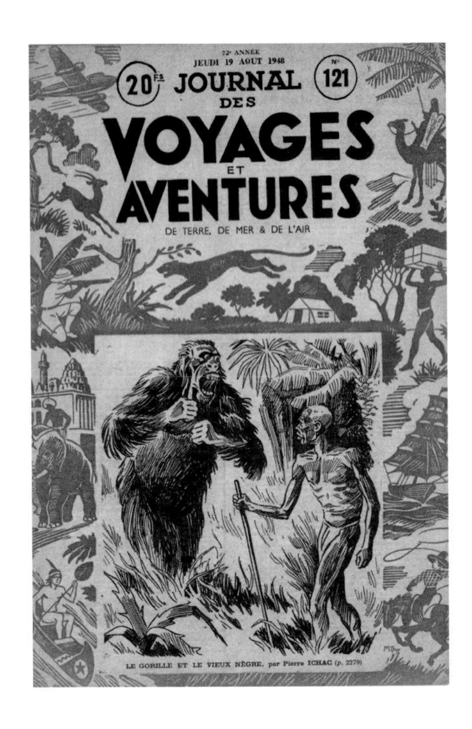

group of Fang people, reported similar sylvan demons that could be killed if the hunters' spears were specially treated by a medicine man.[17] Anthropologists and oral historians continue to report stories today from various parts of central Africa of gorillas inhabited by the spirits of people returned from the dead.[18]

Anthropologists have also recorded stories of gorillas and humans having a common origin or sharing family ties. Philippe Laburthe-Tolra's historical ethnography of the Beti and Bulu peoples of southern Cameroon in the early twentieth century recounts a mythological tale of a gorilla whose mother was human, from an era when gorillas and humans had close ties of kinship and obligation. The gorilla sleeps with one of his uncle's wives and then is refused when he still tries to claim part of his uncle's hunting spoils. Gorillas and other animals co-existing with humans feature too in the historical Ekano or oral legends of the Mpongwe, Benga and Fang peoples, as collected by Robert Nassau in 1914.[19] Du Chaillu described how numerous central African peoples believed that if either a pregnant woman or her husband so much as looked at a gorilla during her term, she would give birth to a gorilla instead of a human child, underlining the gorilla's supposed virility and closeness to humanity.[20]

Secret gorilla cults have been documented among Fang and Bulu groups in Gabon, Cameroon and Equatorial Guinea. In 1863 the Scottish historian William Winwoode Reade wrote excitedly about witnessing a highly physical gorilla dance, performed for him by villagers near the Ngumbi forest in Gabon.[21] He was possibly describing the gorilla-centric *Ngi* cult. The *Ngi* was later to be documented by German ethnographer Günther Tessmann before the colonial authorities, threatened by its wielding of judicial and punitive authority, attempted to suppress it and its adherents around 1910.[22] Tessmann records that the *Ngi* acted as a kind of police force, donning striking elongated masks and

'The Gorilla Dance', wood engraving, from W. Winwood Reade, *Savage Africa* (1864).

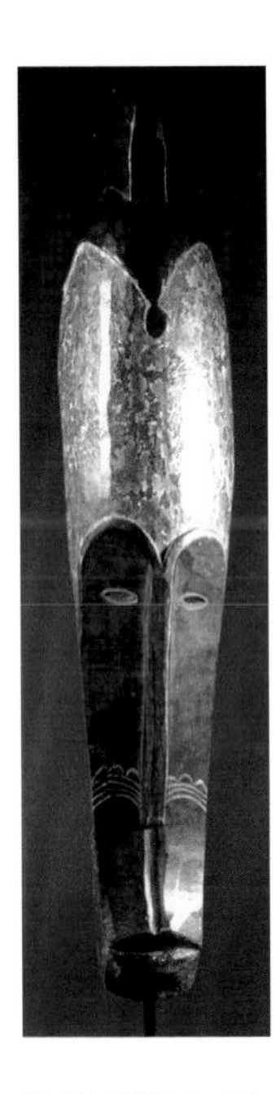

Ngi (gorilla) society mask, Fang people, Gabon.

venturing forth at night to deliver retribution and even death sentences to unruly 'witch doctors' and social recalcitrants. The term *Ngi* signified the gorilla, the arcane cult's symbol of positive energy. It was Fang custom to hang a gorilla skull on the *akôn aba*, the main supporting post at the front entrance to the men's council house (*aba*), where it was at the centre of all ceremonies to do with patrilineage, coming of age and death.[23] Angela Meder has more recently posited the continuing existence of other clandestine gorilla cults in Africa. She also catalogues a range of known practices, such as drinking from gorilla skull vessels or wearing gorilla-teeth necklaces to increase strength, and sniffing dried and ground gorilla bone or inserting it into cuts in the skin for greater power and equilibrium. The sagittal crest of silverback gorillas has been used for carvings by the Fang in Gabon, and masks may be decorated with these crests or with gorilla teeth, skin or hairs, while shields sometimes incorporate gorilla skulls. Dale Peterson has recently documented how gorilla flesh, obtained in the fetish markets of large cities like Brazzaville, is today boiled and ground as a nutritional supplement to empower athletes, or pulverized into an unguent for backaches.[24]

As well as accounts of secretive cults and stories of gorilla transformation, many tales have been told of beliefs about the symbolic power of consuming parts of the gorilla. After Du Chaillu killed his first gorilla while hunting with the Fang, he noted that they ate the entire animal save for the brain, which was reserved for making 'charms of two kinds. Prepared in one way, the charm gave the wearer a strong hand for the hunt, and in another it gave him success with women.'[25] A member of the Percy Sladen Zoological Expedition to the British Cameroons in 1932–3 described how, after a magnificent male gorilla was shot near the Cross River, the chief of the local hunters claimed the right to eat the whole of its small intestine for its 'ju-ju' value. The hunter who had killed the beast took a selection of

> eyelids, armpits, groins, lungs, gut, stomach muscles, and the whole heart and pancreas, showing an amazing knowledge of anatomy. These were all boiled down together with a fowl, half eaten by the hunter, and the rest smeared over the sights of his gun to ensure future kills.[26]

Notwithstanding such prurient interest in the eating of various parts of the animal on the part of Europeans, many central African peoples proscribe the consumption of gorilla meat. Historical anthropologists Tamara Giles-Vernick and Stephanie Rupp record that, according to the oral histories of the Bangando of south-eastern Cameroon, in the nineteenth century, the *dáwá* – their term for a group of animals including gorillas, chimpanzees and monkeys – saved Bangando people from an attack by a Ndzimou group, warning of their approach; in gratitude since that time, the Bangando do not eat the meat of gorillas and other *dáwá*.[27] Meder reports that, at least until the last decades of the twentieth century, the Anyang people from near the Cameroon–Nigeria border used to forbid gorilla hunting on pain of death at all times except as part of a ritual when a new chief was required to eat a gorilla's brain and a high-ranking associate its heart. She also notes that in a number of other areas of Cameroon and of the Republic of Congo where gorillas are hunted, women do not eat gorilla meat.[28] Nevertheless, many forest peoples have routinely included gorilla meat in their subsistence diet. Recent discussion of eating gorilla meat often focuses on the bushmeat (or wild game) trade, which has replaced traditional ritual or subsistence uses of wild animal meat with a highly profitable and illegal trade facilitated by the destruction of forests.[29]

Karl Amman, *African Ape and Gorilla Bushmeat Market: Gorilla Hand*, 1994.

It has been just over a century and a half since the first scientific studies of the gorilla were published in Europe and North America. Before that time, outside of central Africa, parts of the rest of the world had received news of the animal from early travellers' accounts of encounters. There are written accounts dating back millennia of sightings and rumours of human-like animals standing erect. In the fifth century BC Hanno II of Carthage, during a voyage around the western coast of Africa, encountered the 'gorillae' – savage hairy male and female creatures who climbed to safety when attacked, defending themselves with stones. Hanno's men captured, killed and flayed three females; their skins were brought back to Carthage and displayed as trophies in the Temple of Tanit. Numerous other authors in antiquity such as Aristotle (fourth century BC), Pliny the Elder (first century AD) and Galen (second century AD) also wrote accounts of anthropoid or human-behaving hairy creatures inhabiting distant lands.

In the mid-1500s Conrad Gesner's *Historiae animalium* sought to synthesize knowledge of the larger, human-like and tail-less forest-dwelling creatures described by the ancient writers with mythical tales of the hairy wild men (the so-called *Homo sylvestris* or 'forest man') who were reported to lurk on the fringes of the 'civilized' world, marrying his time's notions of imagined wild peoples with genuine, if primitive, observations of primate species. Since, according to biblical authority, god-fearing humans alone lacked a tail and walked upright, any other creature behaving in this fashion was deemed to be morally problematic and threatening.[30]

Purchas, his Pilgrimes, first published by Samuel Purchas in 1614, recounts the adventures of Andrew Battell, an English sailor, captured by the Portuguese in 1559, who spent eighteen years in the Portuguese colonies of Angola and Congo in the 1560s and '70s. Battell described 'two kinds of Monsters, which are very common in these Woods, and very dangerous' – the enormous *Pongo,* who was 'in all proportions like a man, but that he is more like a Giant in stature', and a smaller anthropoid, the *Engeco*.[31] While Battell's 'two kinds of Monsters' are today recognized as the gorilla and the chimpanzee, at the time both creatures were regarded in Europe as the stuff of legends. In 1766 the celebrated French naturalist Georges-Louis Leclerc, Comte de Buffon, published the fourteenth volume of his monumental *Histoire naturelle* in which he argued that there was only one species of African primate, the chimpanzee, which he called the Jocko.[32]

When natural scientists from England, France and the United States encountered the gorilla for the first time in the mid-nineteenth century, it was immediately catapulted to the centre of evolutionary debates on humanity's relationship with primates. There ensued a 'gorilla mania' in Europe and the U.S. for accounts of the gorilla's appearance and behaviour in literature and popular entertainment. This moved into the domain of the fine arts with the display in the Paris Salon of 1859 of Emmanuel Fremiet's

Juste Chevillet after Jacques de Sève, 'Le Jocko', engraving from Georges-Louis Leclerc, Comte de Buffon, *Histoire Naturelle*, vol. XIV (1766).

colossal sculpture of a gorilla carrying off a prone (dead or fainted) African woman, a work that was hugely influential as the first European artistic response to scientific accounts of the animal. Despite this sculpture having the title *Gorille femelle* (female gorilla) chiselled into its pedestal, the poet Charles Baudelaire and others who viewed it at the Salon considered this a smokescreen intended to obscure what they felt was clearly a sexual scene. This work firmly established a trajectory of images of gorillas as lustful beasts sexually interested in human women that leads down through Klimt to King Kong and beyond.

The artistic, literary and cinematic mania for the gorilla which would grow exponentially from the time of Fremiet's *Gorille femelle* onwards did not necessarily feel itself beholden to an appearance of realism or authenticity. For more than a generation from the early 1920s, cinema-goers' most frequent sighting of a gorilla actually involved a succession of furry-suited professional gorilla actors. The most famous of these remains Charles Gemora, a Philippines-born make-up artist who worked a profitable sideline as a specialist primate impersonator. Gemora, who made his own gorilla suits and studied the animals in San Diego

Emmanuel Fremiet, *Female Gorilla Carrying off a Negress*, 1859, life-size plaster (destroyed).

Zoo for inspiration, starred as a gorilla in films alongside Bela Lugosi, Bob Hope, the Marx Brothers, Lon Chaney, Abbot and Costello, Laurel and Hardy, Bing Crosby and Robert Mitchum, among other Hollywood greats. While seeming at first limited, Gemora's repertoire grew to be both varied and nuanced, ranging from Erik the perverted sex-fiend ape in Robert Florey's *Murders in the Rue Morgue* (1932) to the noble Gibraltar in the Marx Brothers comedy classic *At the Circus* (1939), in which he performed cinema's first gorilla trapeze act. By the 1940s Gemora was reportedly earning $5,000 a picture for his gorilla impersonations, delighting audiences who did not seem to mind at all that he walked comfortably on two feet human-style and made a variety of imaginative snuffling sounds.[33]

Gemora was followed by a succession of ape actors who worked across the Hollywood studios, frequently playing a gorilla uncredited for one company while gainfully employed in more standard roles by another. These included Ray Corrigan and Emil van Horne in the 1940s, and Steve Calvert and George Barrows

Gorilla impersonator Charles Gemora with Lou Costello in *Africa Screams* (1949).

George Barrows, gorilla-suited as Ro-Man in *Robot Monster* (1953). Courtesy Wade Williams Distribution.

in the 1950s. Their lives have been documented by one of their own, the more recent gorilla impersonator Bob Burns.[34] These cinema gorillas invited their viewers on a journey of happily suspended belief, at times wickedly sending up gorilla history – as, for example,with the male gorilla in Paramount's *Road to Bali* (1952) who bends a rifle in half like warm toffee in a playful echo of the original illustrations to Paul du Chaillu's *Explorations and Adventures in Equatorial Africa*; or his amorous gorilla wife (played by Steve Calvert) who tries to run off with both Bob Hope and Bing Crosby in a reversal of the sexed-up male gorilla stereotype.

Steve Calvert and Ray Corrigan, gorilla-suited in *Bela Lugosi Meets a Brooklyn Gorilla* (1952). Courtesy Wade Williams Distribution.

Bob Hope and amorous gorilla in *The Road to Bali* (1952).

J. Whitney, 'Death of My Hunter', wood engraving, from Paul du Chaillu, *Explorations and Adventures in Equatorial Africa* (1861).

While these performances may seem laboured or camp from today's perspective, they enchanted their contemporary viewers. Concerning Gemora's role as Gibraltar in *At the Circus*, Groucho Marx recalled: 'At the first preview nobody mentioned us. Our gifted performances went for naught. The audience had eyes only for the gorilla.'[35]

2 Good Shot: Europe and America Meet the Gorilla

> Guns have metamorphosed into cameras in this earnest comedy, the ecological safari, because nature has ceased to be what it had always been – what people needed protection from. Now nature – tamed, endangered, mortal – needs to be protected from people. When we are afraid, we shoot. But when we are nostalgic, we take pictures.
> Susan Sontag, *On Photography* (1977)[1]

Europe and North America received confirmation of the long-rumoured existence of the gorilla only in December 1847, when the *Boston Journal of Natural History* published an article by Protestant clergyman, missionary and physician Thomas Staughton Savage and his colleague Jeffries Wyman of Harvard University. Savage had been shown a strange skull in April 1847 in Gabon at the home of his friend Reverend John Leighton Wilson, Senior Missionary of the American Board of Commissioners for Foreign Missions to West Africa, who told him how it was 'represented by the natives to be that of a monkey-like animal, remarkable for its size, ferocity and habits'.[2] Savage proceeded to identify this new ape with the *Pongo* described by sailor Andrew Battell in the late sixteenth century, and the *Ingena* that the English adventurer Thomas Bowditch had heard rumours of in 1819, noting that it was called the *Engé-ena* by the Mpongwe people who lived around the banks of the Gaboon River. He distinguished it firmly from the *Enché-eko*, the Mpongwe name for the chimpanzee which also dwelled in the same region, noting that while the *Enché-eko* was to be found closer to the West African seaboard and hence was a known if long misunderstood primate, the *Engé-ena* kept to the interior of Lower Guinea, which accounted for its rarer sighting and the numerous myths and legends that had grown up around it. Savage and Wyman chose the name *Gorilla* for their newly described primate from the ancient account of hairy creatures seen by Hanno II of Carthage in the fifth century BC. Their article was accompanied by four accurate lithographic studies by W. H. Tappan of a male and female gorilla skull, the first images of the gorilla to appear outside of Africa. The animal remains studied and drawn for this landmark publication were specimens of what is now referred to as the Western lowland gorilla.

W. H. Tappan, '*Troglodytes gorilla* (Savage). Head of Female, Front View', lithograph, published in *Boston Journal of Natural History* (December 1847).

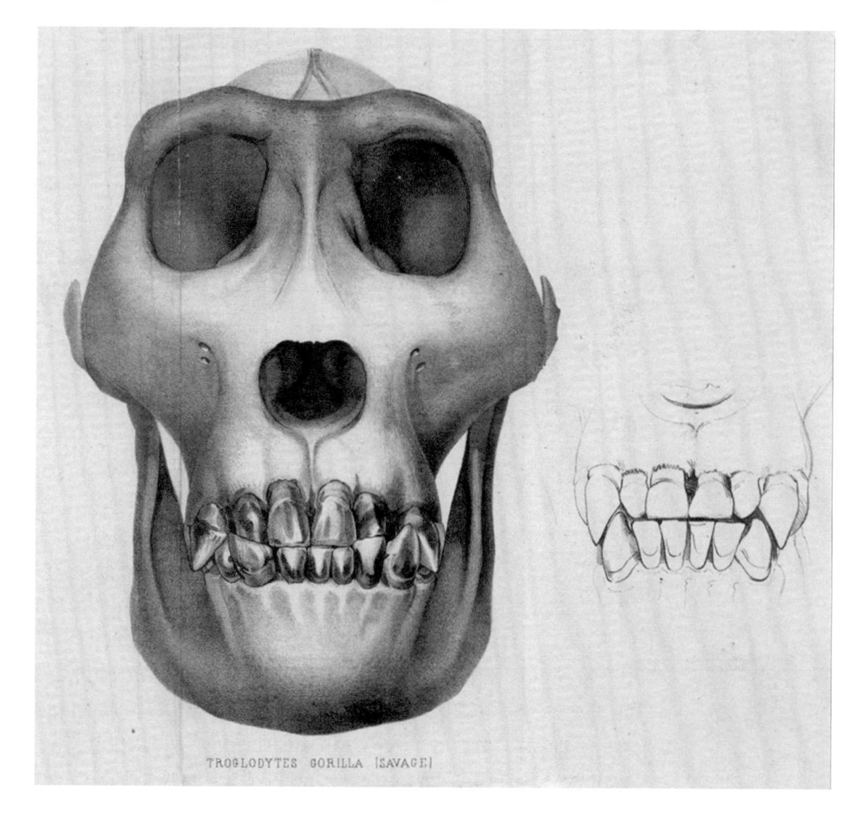

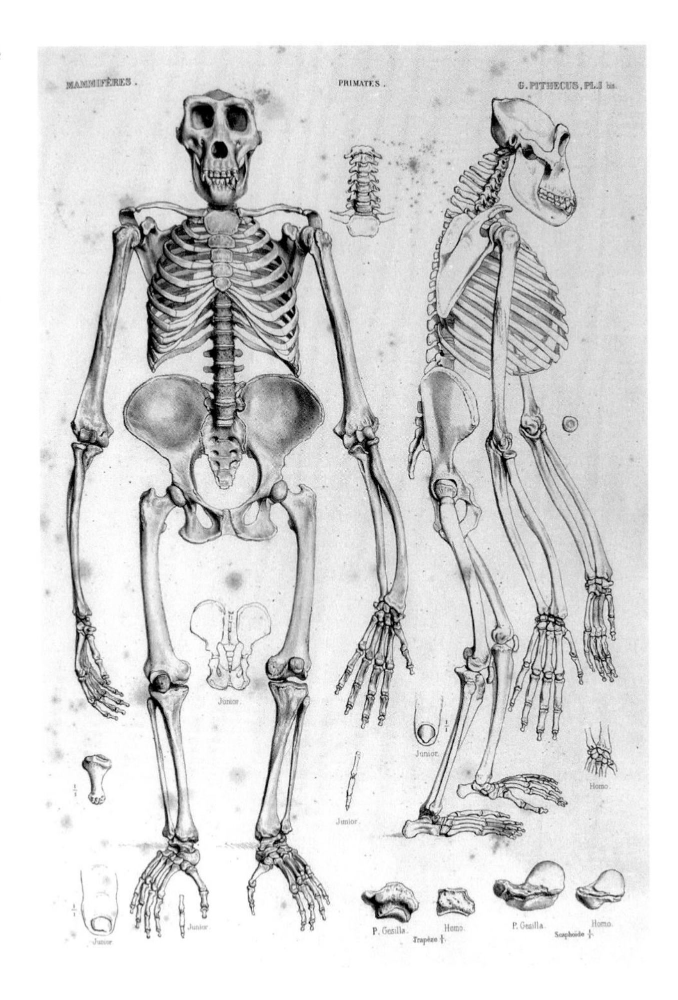

Jacques-Christophe Werner or Emmanuel Fremiet, 'Skeleton of Female Gorilla Gina (Savage)', lithograph, from Henri-Marie Ducrotay de Blainville, *Ostéographie-Atlas*, vol. IV (1849).

There ensued a vigorous trade in bottled, boned and stuffed gorillas between Africa, Europe and North America as museums and scientists engaged in a scramble to acquire specimens of this new species of primate, which was to play a central part in the nascent evolutionary debates. Scientific study of the gorilla began early in France, which benefited from trade agreements it had established with Gabon in 1839. The first complete gorilla specimen to arrive in France, the full skeleton of an adult female collected in Gabon, was offered to the Muséum d'Histoire Naturelle in Paris in April 1849 by a Monsieur Gauthier Laboulaye, a surgeon working for the French navy. A lithograph after this skeleton signed by the Muséum's scientific illustrator Jacques-Christophe Werner, the first gorilla image to appear in France, was published in that same year in a supplementary fascicle of the palaeontologist Henri-Marie Ducrotay de Blainville's *Ostéographie*, a comparative study of the skeletons of fossilized and living vertebrates. Two years later the Muséum received the first two whole gorilla bodies to be seen in Europe. These were an infant (boarded as a live passenger, then embalmed after its death in transit)

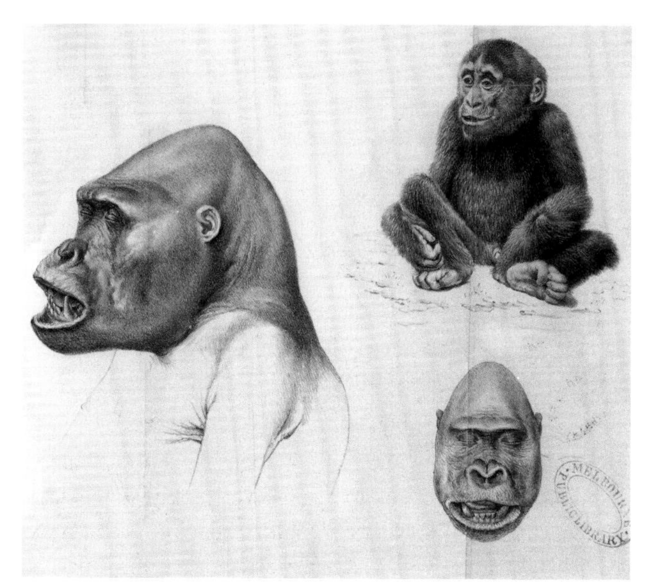

'Gorilla (troglodytes gorilla savage) (after the Museum's daguerreotypes)', lithograph, published in *Annales des sciences naturelles*: *Zoologie et biologie animale*, 3rd ser., vol. xvi (1851), artist unknown.

and a full-grown male that had been secured by a naval doctor, Monsieur Franquet, preserved in alcohol and shipped back from Gabon by Captain Penaud of the *Eldorado*. On 19 January 1852 Isidore Geoffroy Saint-Hilaire, Professor of Zoology (Mammals and Birds) at the Muséum, delivered the first paper on these preserved gorillas to the Académie des Sciences in Paris. He noted that Franquet had measured the adult male, prior to its steeping in alcohol, as reaching to 1.67 metres in height, a grandeur of stature that he illustrated dramatically with a life-size profile study of the new animal drawn by Werner, as well as photographic studies of the preserved specimens taken by the Muséum's chemical preparator, Monsieur Terreil.[3] Once taxidermied by Monsieur Portmann, the Muséum's zoological preparator and model-maker, this adult male specimen occupied a transparent showcase in the centre of the Ape Room in the Galeries de Zoologie. Nearby, visitors could view a daguerreotype of the gorilla as it first arrived, squashed out of shape in a vat of alcohol, 'so as to better appreciate the tremendous skill with which Monsieur Portmann has mounted up this colossal creature'.[4]

Across the Channel Richard Owen, Hunterian Lecturer in Comparative Anatomy at London's Royal College of Surgeons and (from 1856) Superintendent of Natural History at the British Museum, was also keen to obtain examples of the new primate, *Troglodytes gorilla (Savage)*. In the 1850s the skeleton of an adult gorilla, acquired in 1851, was a prize exhibit in the British Museum's Mammalia Saloon, where it dominated nature's other examples of the anthropoid (human-like) apes, such as the orang-utan and the chimpanzee. A photograph by Roger Fenton that compared this skeleton with that of an adult human was a popular postcard image in the Victorian era. Doubtless jealous of the Paris Muséum's gorillas, Owen naturally wanted to acquire a complete gorilla specimen himself.

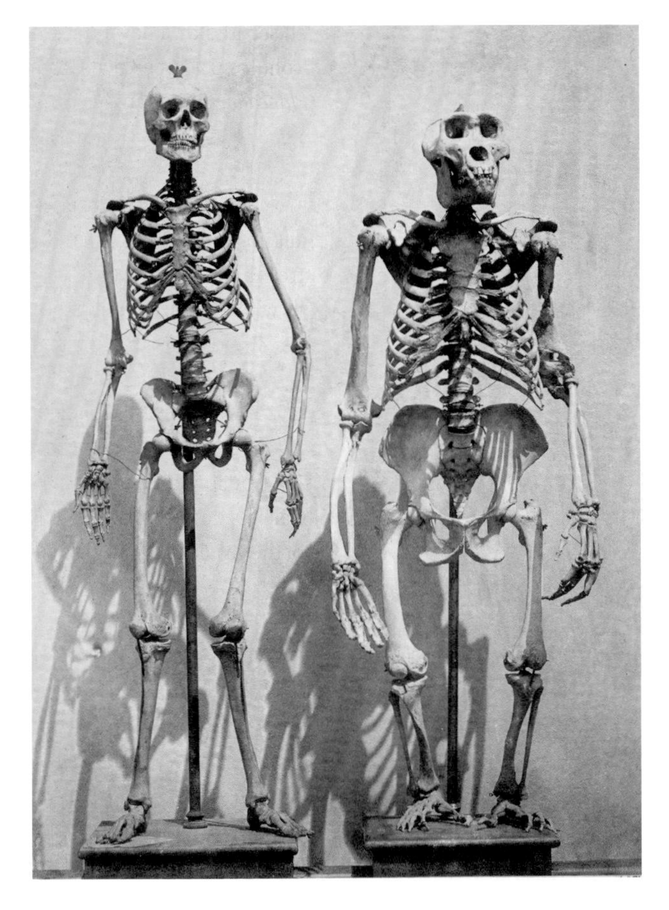

Roger Fenton, *Skeleton of Man and of the Male Gorilla (Troglodytes Gorilla) I*, 1854–8, salted paper print.

The arrival in September 1858 of Owen's first complete gorilla import was something of a disaster. 'Upon the opening of the cask', reported the *Leisure Hour* magazine,

> the appearance of the specimen was such as to leave little hope of its being saved, and the stench was so great that it

was found necessary to have the cask immediately closed [because] the animal was in such a state of decomposition.[5]

Smothering his disappointment, Owen had these decomposed remains photographed by Peter Ashton before turning them over to Abraham Dee Bartlett, a pioneer in the field of taxidermy and future Superintendent of the London Zoo, who had also been present at their uncrating. It was Bartlett who eventually brought some semblance of life to Owen's decaying gorilla, providing Britain with its first realistic glimpse of this feared new animal, which the *Illustrated London News* in 1859 noted was

> found only in the thickest jungles of Western Africa, and is so wary, active and fierce that a close inspection is almost

C. P. Nickolls (after Peter Ashton), 'The Gorilla, as Photographed in the Tub in which It Arrived', wood engraving, published in *The Leisure Hour* (27 January 1859).

'Specimen of the Gorilla as Restored at the British Museum', wood engraving, published in the *Illustrated Times* (1859).

> an impossibility, especially as the moment it sees a man it attacks him . . . Sheer malignity prompts the animal to this course, for it does not eat the dead man's flesh, but finds a fiendish gratification in the mere act of killing.[6]

As previously mentioned, at the Paris Salon of 1859 the French sculptor Emmanuel Fremiet, who had studied under Werner at the Paris Muséum, unveiled the world's first life-size sculpture of a gorilla. Rejected by the Salon jury as 'seriously offensive to public morality', but defended by the Minister for the Arts, Fremiet's *Gorille femelle* was installed towards the entrance of the Salon's sculpture section, set back in a niche behind heavy green velvet curtains.[7] Here it became a cause célèbre, shocking and titillating

'The Ferocious Gorilla', wood engraving, published in the *Illustrated London News* (9 April 1859).

Parisians in equal measure. Fremiet's sculpture, which depicted a gorilla dragging an apparently dead African woman off to its lair, may have been influenced by the accounts of Gauthier Laboulaye who, when presenting the Muséum with its first gorilla skeleton in 1849, had passed on tales of gorillas 'slaking their lust by ravishing Negro women wandering carelessly in the woods'.[8] While Fremiet's gorilla was a female, its anatomy and morphology doubtless having been based upon close study of Gauthier

Laboulaye's female skeleton, it was read at the time as a male ape, caught in the act of sexual ravishment. This compounded fantasies of the gorilla as a sexual predator for audiences already familiar with the brutal sex-crime butchery committed by the murderous orang-utan in Edgar Allan Poe's *Murders in the Rue Morgue* of 1841 (which in nineteenth-century illustrations of the volume was often depicted more as gorilla than orang-utan, as for example by Aubrey Beardsley; this is also true of later cinematic adaptations).

Accounts of the gorilla's appearance and behaviour for European and American audiences moved into another register with the lectures and expeditionary writings of Paul du Chaillu, the French/American adventurer who became the first European to observe and hunt gorillas in the wild. As a teenager Du Chaillu had lived for a time in Gabon with Revd John Wilson, close to the time when Wilson first told Savage about the strange new animal skull that became *Troglodytes gorilla (Savage)*. After migrating to the U.S. in 1852, where he initially worked as a French teacher, Du

Emmanuel Fremiet, *Female Gorilla Carrying Off a Negress*, 1859, life-size plaster (destroyed in 1861).

J. Whitney, 'Death of the Gorilla', wood engraving, from Paul du Chaillu, *Explorations and Adventures in Equatorial Africa* (1861).

Chaillu soon became a popular speaker about his experiences in Africa. Sponsored by members of the Academy of Natural Sciences of Philadelphia, he returned to Gabon in 1855 on a major collecting expedition.

Du Chaillu came back to the U.S. in late 1859 armed with thrilling gorilla stories and a travelling exhibit of taxidermied specimens to 'prove' them. He was soon asked by *Harper's Magazine* to prepare his tales for publication, and invited to share his adventures with audiences in the United Kingdom. At his inaugural address to the Royal Geographical Society in February 1861, Du Chaillu amazed a capacity crowd with his tales of shooting Western lowland gorillas in Gabon, and his accounts of their natural habitat and strange behaviour – such as the blood-curdling vocalizations and breast-beating of male gorillas when agitated. He also dazzled his audience with the enormous skeletons, skins and stuffed gorillas he displayed to illustrate his lectures. Richard Owen acquired the best of Du Chaillu's specimens for the British Museum, which quickly placed them on display and extended its opening hours in expectation of increased visitor numbers.[9]

This landmark lecture was followed by the publication of Du Chaillu's adventures in simultaneous American and British editions in May 1861. *Explorations and Adventures in Equatorial Africa* was a runaway success, selling more than 10,000 copies in the first two years alone.[10] It was the passages in Du Chaillu's book dealing with his face-to-face slaying of ferocious gorillas that caught the public's imagination. The gorilla was a 'monster of whose ferocity, strength and cunning the natives had told me so much', he thundered, 'an animal scarce known to the civilized world, and which no white man before had hunted'. It had 'a hellish expression of face, which seemed to me to be like some nightmare vision . . . some hellish dream creature', and its yells of rage 'remind one only of the inarticulate ravings of a maniac'. Du Chaillu was able to touch the hearts of his readers, even while terrifying them, in describing his mixed emotions: 'I protest I almost felt like a murderer when I saw the gorilla this first time . . . they looked fearfully like hairy men . . . running for their lives.'[11] Everyone was talking about his book, whether they had read it or not.

Attacks on Du Chaillu from jealous rivals attended his success, and in a media frenzy known as the 'Gorilla War' he was accused of everything from fabricating his stories to shooting his gorillas in the back or never having shot one at all, buying them instead from other hunters. Having been London's 'Lion of the Season' for a good part of 1861, Du Chaillu saw his reputation reach a sudden nadir when, exasperated beyond self-control, he menaced and spat on one of his detractors at a meeting of the Ethnological Society. *Punch* remarked dryly: 'Such a way of arguing may be tolerated possibly at a meeting of Gorillas, but, happily, among Englishmen it has not yet been sanctioned.'[12]

Stung by criticism and scepticism concerning his findings, Paul du Chaillu returned to Gabon in 1863–5, this time carrying photographic and other scientific recording equipment in order to prove beyond doubt that his first accounts of discovering gorillas had been entirely truthful. The results of this second expedition were published in his *A Journey to Ashango-Land* in 1867. In preparation, Du Chaillu studied photographic techniques with Antoine Claudet, a Frenchman who had established one of the first commercial photography businesses in London.[13]

During this second foray into Gabon, Du Chaillu photographed gorillas as they were tethered up at his base camp after their capture in the nearby jungle. Unfortunately, he recounts that mid-way through this second trip, most of his photographic equipment was stolen by members of the Ashira tribe, and that during a later retreat from the Mouaou people, while the expedition was under fire from poisoned arrows (one of which slightly wounded Du Chaillu), all of the photographs were also lost. Nonetheless, Paul du Chaillu seems to have been the first person to photograph the gorilla in the wild, even though the results did not survive his second journey to West Africa. Due to the very limited knowledge of the care and handling of primates in captivity at this time, nor did his gorilla subjects survive more than a few days after capture. Only one live gorilla was shipped to London from this trip, which died during the sea voyage.

Du Chaillu also shot gorillas on commission during this return trip to Africa, providing countries as far away as Australia with gorilla exhibits for the instruction and entertainment of local audiences who had been following the 'Gorilla War' in London. Frederick McCoy, Professor of Natural Science at the University of Melbourne and Director of the National Museum of Victoria, wrote to London as early as 24 August 1861, immediately after the publication of Du Chaillu's first African narratives, seeking to acquire some of his gorilla specimens.[14] McCoy authorized the purchase for Melbourne of three of Paul du Chaillu's gorillas – a

'Photographing Gorillas', wood engraving, from Paul du Chaillu, *The Country of the Dwarves: A New Work of Stirring Adventure* (1872).

'grandfather', 'mother' and infant – slain during the early days of Du Chaillu's second expedition to Gabon. The installation of this new gorilla group at the Melbourne Museum in late June 1865 provided both an intellectual coup and a crowd-pleasing exhibit for McCoy's institution.

Europeans enamoured of big game hunting for sport quickly focused their energies on the gorilla, largely through the influence of Robert Ballantyne's *The Gorilla Hunters*, a popular children's book of 1861 based upon Du Chaillu's accounts of gorilla hunting. After shooting his first gorilla, Ballantyne's narrator, Ralph Rover, tells his young readers that

> It is impossible to convey in words an idea of the gush of mingled feelings that filled our breasts as we stood beside and gazed at the huge carcass of our victim. Pity at first predominated in my heart, then I felt like an accomplice to a murder, and then an exulting sensation of joy at having obtained a specimen of one of the rarest animals in the world overwhelmed every other feeling.[15]

Oswald Rose Wilson, 'The Gorillas in the Melbourne Museum', wood engraving, published in *The Australian Museum for Home Readers* (25 July 1865).

This book remained in print for juvenile readers for several generations, the graphic treatment of respective editions frequently associating the gorilla with its nemesis, the gun. In 1894 Henry Bailey published his account of big game shooting in the Congo, including his bagging of a female gorilla:

> On perceiving me she immediately stood up on her hind legs facing me, and, quicker than it takes me to write this, I had covered her with my rifle, sending a bullet through her chest, the shot passing through the top part of her heart and dropping her dead without a sound.[16]

By the turn of the century 'big game hunting' postcards featuring gorillas and European shooters had become commonplace. A typical example from 1909 describes the gorilla as a 'brutal looking animal' of 'colossal strength, that it uses in its native habitats against men'. The equation was simple – the only appropriate response when encountering a gorilla was to reach for one's gun.

Cover of R. M. Ballantyne, *The Gorilla Hunters* (1930 edition).

In 1909 the rogue German soldier, hunter and journalist Fritz Duquesne wrote about dangerous gorilla hunts, as did Colonel Etienne Bazin, who argued that the ferocity of the beast was such that 'a gorilla does not pay the lightest heed to a high velocity, small caliber bullet, even through the heart at times'. Exaggerated as their stories doubtless were, they were at least grounded in some truth, unlike the wild tales that peppered the American press in the 1880s and '90s of gorillas battling lions and even shooting rifles back at intrepid naturalists.[17] Whether true or

Big Game Hunting, postcard inspired by Theodore Roosevelt's safari to Africa, Chicago, 1909.

Shooting a Gorilla, McLaughlin's Coffee trade card, Chicago, *c*. 1893.

false, such tales made the gorilla seem like the big game hunter's ultimate prize. The toll exacted upon the Western lowland gorilla as a result of such tales will probably never be known.

While climbing Mt Sabyinyo in the Virunga Volcano ranges in 1902, the German officer Robert 'Oskar' von Beringe recalled how 'we spotted from our camp a group of large, black apes, which attempted to climb to the highest peak of the volcano'.[18] Von Beringe's party shot two of these gorillas, and he sent a skeleton to an anatomist colleague at the Zoological Museum in Berlin, Paul Matschie, who classified the creature as a new species, calling it *Gorilla beringei*, the mountain gorilla. Later this classification was revised to that of a sub-species, *G. gorilla beringei*.[19] The mountain gorilla differs from its lowland kin primarily by possession of a thicker and shaggier dark coat, a necessity for its survival on the cold and wet high slopes of the Virunga Volcanoes. News of the discovery sparked a new wave of hunting, with the result that between 1902 and 1925 some 54 *G. g. beringei* were shot and exported from the Virunga Volcanoes.[20]

Just as Herr Paschen's gorilla became the most celebrated specimen of its kind in 1901, the famous photograph of its corpse propped up by local bearers set the tone for how gorillas were to be imaged for decades to come. Dead gorillas litter the pages of scientific and popular literature from the 1920s and '30s – posed either seated like an overstuffed Steiff teddy bear (often juxtaposed with Africans of lesser stature, strategically placed for comparative purposes), or strung up crucifixion-style, hung by the wrists from makeshift scaffolds. The latter type of 'crucified' gorilla was the frontispiece that veteran adventurer Sir Harry Johnston thought 'illustrated so splendidly' Thomas Alexander Barns's *The Wonderland of Eastern Congo* (1922), an account of Barns's shooting of gorillas, hippopotami, elephants, buffalo and lions in and around the Virunga Volcanoes.[21]

Among the 1,000 mammals, 1,700 birds and 10,000 insects that Prince William of Sweden and his team killed on the Swedish Zoological Expedition to Central Africa in 1921 were fourteen mountain gorillas, two from each of the Virunga Volcanoes. He had received special authorization from the Belgian state to shoot such a large number of the mountain gorillas, for 'the animals are otherwise under protection'. The prince's passion for killing was matched by his pronounced racism (he described one native bearer as 'not unlike an improved edition of a gorilla'), a trait that permeates most travel writing about Africa by Europeans at this time. Despite the relative ease with which his guns felled animal after animal, the prince's impressions of the gorilla, informed by its reputation as a fearsome monster, as well as the physical difficulties posed by the Virunga terrain, led him to conclude that 'Gorilla hunting is, in my opinion, one of the most difficult and most fatiguing sports in existence, but when you succeed the gorilla is a worthy object.'[22]

Charged with shooting his own mountain gorillas for the British Museum, Thomas Barns nonetheless criticized Prince William of Sweden for his excessive cull of fourteen of the animals. Despite lamenting that 'with the opening up of Africa . . . the unsuspicious gorilla will be the first to go, if strict game laws are not enforced for his protection', and his professed misgivings that 'no one with a spark of feeling can free himself from the thought that killing them is akin to murder', trophy photographs of dead gorillas featured heavily in Barns's African travel books.[23] Barns's sympathy for the endangered gorilla was also to prove short-lived, as by the mid-1920s he was attempting to set up his own African 'adventure tours' business which would offer the gorilla on its menu for prospective big-game hunting clients.

In 1926 Barns was to write, in response to press rumours of the gorilla's demise, that

Mountain gorillas, Ntambara Group, 8 April 2009.

> these animals were not so rare as supposed . . . that gorillas existed in the Western Kivu forests in great numbers and that all this outcry of the impending extermination of the gorillas was exaggerated.

He then provided observations on 'the gorilla solely as a sporting adversary, and of the danger attached to hunting him . . . so that if any of my readers contemplate gorilla hunting, they will, at least, know what to expect'.[24] Two years earlier, convinced that gorillas numbered in the thousands, Barns had suggested 'that one or two gorillas be allowed to be shot on the 2000 f. license already in force, which would induce many sportsmen to visit the country'.[25]

In 1929 the Henry C. Raven expedition from Columbia University and the American Museum of Natural History travelled to the Belgian Congo and the French Cameroons in search of more gorillas for scientific and medical research. The Raven expedition had a special mission to shoot gorillas only in the head, causing minimal damage to the blood vessels so that their bodies could be embalmed evenly and completely and shipped whole back to America. Despite opposition from the Belgian ambassador in London, who had 'become an ardent conservationist to the point of being unwilling to allow a single gorilla to be killed for even scientific purposes', the expedition obtained

'The rare Kivu Gorilla, shot by the Author on the Virunga Mountains, and the boy Salim', from T. Alexander Barns, *The Wonderland of the Eastern Congo* (1922).

its permits, shot its quota of five apes and subsequently published a plethora of photographs of the fallen primates posed with local bearers and the triumphant white hunter, gun in hand.[26] It fell to Raven, a solitary hunter by preference, to make the five clean kills that the expedition needed. His view was that 'Although it was indeed a pity to kill so noble a living monument of past ages, we had not murdered him wantonly and for sport.'[27] It was further claimed that 'the eating of men being out of date, the natives find the gorilla a good substitute . . . perhaps ten times as many gorillas are killed by the natives themselves as are killed by white men'.[28] To his credit, Raven did resist offers to organize mass gorilla drives involving dozens of villagers, taking instead nine months to cleanly kill his five lowland gorillas. For this expedition Raven packed, among other guns, a 30-30 calibre Savage rifle. The Savage Arms Company, founded in Utica, New York, in 1894, had recently been devoting advertising copy to the gorilla, 'the huge, silent, uncanny man-monkey who glides through the dark, tangled African jungle as swiftly and easily as a shadow', and threats of 'his supernatural vitality, his fiendish intelligence and his horrible vindictiveness when attacked', to promote sales of their rifles as a solution to 'the most unnerving problem a hunter can face'.

Slaughtered gorillas displayed crucifixion-style also featured in the writings of the Italian Commander Attilio Gatti, who was issued permits by the Belgian authorities to hunt gorillas for university collections in Florence and Johannesburg in the early 1930s. One of these gruesome images, of a gorilla hung by the wrists from a rudimentary bush scaffold, adorned the dust-jacket of Gatti's monograph *The King of the Gorillas* of 1932, and others appeared in the pages of articles that Gatti placed in youth-oriented journals such as *Popular Mechanics* and *Boys' Life*. Here he boasted of his bravery in besting 'the ferocious gorilla, a very dangerous beast, probably the most dangerous of all African animals', in company with the Mambuti (M'buti) pygmies in the Tchibinda forest.[29] Tellingly, Gatti's account of shooting gorillas was paired in the issue of *Boys' Life* for May 1932 with an advertisement for Winchester .22 rifles. After a subsequent visit to the Belgian Congo to film gorillas, during which he was warned that his shooting of more of these animals, even in self-defence, would result in prosecution, Gatti ranted against what he perceived to be the twin myths of gorillas being both totally gentle and threatened with imminent extinction, arguing the case for the 'hundreds of zoos and museums of every country which have hitherto been unable to secure a specimen of gorilla'.[30]

The extent to which gorillas were shot prior to the Second World War in the name of science and museology can be guessed at from the memoirs of George Merfield, who hunted the apes professionally in the Cameroons in the 1920s. He states that in the four years he spent amongst the Mendjim Mey people, he collected 115 dead gorillas for European museums.

While expressing a distaste for 'idle fatheads whose only wish is to acquire a spurious glory by the slaughter of some of Africa's finest animals', Merfield also admitted to having at times organized gorilla hunts for big game sportsmen. Merfield justified his tracking of a kill for a fatuous European manufacturer by picking out a lame and solitary silverback who appeared ready for death anyway. He was proud, however, to accompany Major P.H.G. Powell-Cotton, the renowned naturalist and big game hunter, on gorilla-shooting expeditions at the end of the 1920s.[31]

Merfield also described the savage slaughter that attended the gorilla round-ups organized by local peoples living near the Cameroon forests as punishment for the damage done to plantain crops by the apes (who ate the pith of the stems rather than the fruit, thereby causing considerable damage to plantings).

The Mendjim people, Merfield recounted, hunted the gorilla for meat, although its flesh was forbidden to women; in the village of Arteck 'there were many gorilla skulls and bones to be seen, and most of the men wore belts of gorilla skin'. During these hunts Merfield frequently observed the gorilla's familial bonds that George Schaller was to document so fully at the close of the 1950s. 'Gorillas never abandon their wounded until they are forced to do so,' he noted, 'and I have often seen the Old Man trying to get a disabled member of his family away to safety.'[32]

The ubiquity with which the visual linkage of the gorilla and the gun permeated popular culture in the U.S. can be gauged from children's toys, such as a 1940s 'Roaring Gorilla Shooting Gallery', a battery-powered combination of tin gorilla and gun in which the gorilla flashed its eyes, raised its arms and emitted a roar every time it was shot in the stomach. In 1964 Hanna-Barbera released a new televised cartoon series, *Magilla Gorilla*. This featured an orphaned ape raised in Peebles Pet Store. In each episode the store owner, Mr Peebles, attempts in vain to sell Magilla, who is returned by client after client following amiable unruly antics. In the first episode of the series, 'The Big Game', the billionaire J. Wimple-Dimple purchases Magilla, who then finds that he has been acquired in order to feed Wimple-Dimple's passion for wild game hunting. A cat-and-mouse game follows, parodying the predation of gorillas in the wild by gun-toting huntsmen, from which Magilla escapes, returning triumphantly to Mr Peebles. The gentle fantasy of *Magilla Gorilla* contrasts with the brutal advertisements that Sturm Ruger were running at this time in magazines such as *American Rifleman*, which depicted hunter Frank Delano brandishing his Ruger .44 Magnum Carbine next to his gape-mouthed kill, a lone gorilla – 'very dangerous and intelligent / fast as a leopard / strong as 14 men / with a terrible cry' – once again strung up crucifixion-style. Small wonder that in early scenes of the alternative history film *Planet of the Apes* (1968), Charlton Heston and his astronaut colleagues witness human scarecrows splayed like crucifixion victims and grinning gorilla guards photographing themselves with their fresh human kills.

One of the first people to actively campaign for the preservation of the gorilla was a man who had initially set out to kill them in earnest: Carl Akeley. A taxidermist who had perfected his craft for client institutions like Chicago's Field Museum of Natural History and New York's American Museum of Natural History, Carl Akeley travelled to the Belgian Congo in 1921 on a specific gorilla-seeking mission with his friends and financial backers Mr and Mrs H. E. Bradley, their five-year-old daughter, her governess and Akeley's secretary Martha Miller. His express purpose in bringing white women along, he claimed, was to make hunting gorillas less attractive as a virile sport by showing that it was not dangerous.[33] Akeley's main mission was to shoot a group of gorillas for mounting in his dreamed-of African Hall at New York's American Museum of Natural History. At this time, as Mrs Bradley noted in her memoirs of the expedition, there was

> no gorilla in a museum mounted by a man who ever saw a wild gorilla. Skins have been bought from hunters and collectors and stuffed according to the best available information. Almost nothing of the animal's habits or capacities has been discovered.[34]

Akeley shot a male gorilla on the party's first foray into gorilla country, within minutes of his first sight of the animal in the wild. The next outing yielded a dead female and infant, thus ensuring a family group of specimens, so that on the third foray he shot instead with the lightweight and flexible camera he had specially designed for the purpose, which he had nicknamed appropriately

'The Gorilla'. With this apparatus he now captured on film for the first time in the wild a mother, baby and small group of other gorillas. 'Almost before I knew it,' he recalled,

> I was turning the crank of the camera on two gorillas in full view with a beautiful setting behind them. I do not think at

The Lone Male of Karisimbi, shot by Mr Bradley, 1921.

> the time that I appreciated the fact that I was doing a thing that had never been done before.[35]

This may have helped diminish the guilt he had felt after his local guides had speared an infant gorilla, which he reached before its death to find that: 'There was a heartbreaking look of piteous appeal on his face which I shall never forget. Like a little child sick and lost, he needed help. I knew that he would have come to my arms for comfort.'[36] In his memoir *In Brightest Africa* (1923) Akeley wrote: 'I was the savage and the aggressor', and that 'much of the shooting that I have had to do in order to obtain specimens for museum collections . . . has made me feel a great deal like a murderer'.[37]

Akeley had begun to grow tired of killing even before setting out on this gorilla expedition. 'One of the big objects of the whole expedition is to kill the idea that hunting gorillas is to be considered a sport', he argued in July 1921. 'Everyone is getting mad to hunt gorillas and I hope that I will be able to influence legislation to prevent the hunting of gorillas except for strictly scientific purposes.'[38] Despite having a licence to shoot ten gorillas, Akeley's party killed only five. Akeley returned to the U.S. with these impressive specimens, which he mounted over the following five years for the gorilla diorama in his projected African Hall, work on which was only fully completed in 1936, long after Akeley's untimely death due to fever and exhaustion on his return to the mountain gorilla's habitats in 1926. Most importantly, though, he was able to argue strongly – and from first-hand observation of gorillas in the wild – against the stories of the aggressive behaviour of these primates which from the time of Paul du Chaillu onwards had been used as an excuse to justify their continued hunting as sport game; indeed Akeley suggests that Du Chaillu was asked to rewrite his accounts until his publishers were satisfied with the animals' ferocity. 'All I want to point out', Akeley wrote, 'is that the gorilla should be judged by what he does, not by how the people who hunt him feel.' Akeley also stressed emphatically that

> The gorilla is a vegetarian, so he kills no animals for food, and he has not progressed sufficiently along the paths of man to enjoy killing as a sport. He lives in amity with the elephants, buffalo, and all the wild creatures of his neighbourhood.[39]

In addition to having shot the first moving images of gorillas in their native habitat, Akeley also became the first taxidermist to have prepared gorilla displays after observing these creatures in the wild.[40] He made death masks of each gorilla that he killed as well as photographing them, a careful recording 'that enables me to make the face of each gorilla . . . a portrait of an individual'.[41] His empathy with the gorilla also led him to make bronze depictions of the apes, using art to reinforce his message about their essential 'humanity'. It has been argued that 'in his description of his gorilla encounters, Akeley endowed the gorillas with features and emotions reminiscent of human family interactions to reinforce notions of human-gorilla similarity'.[42] His sculpture *The Chrysalis* went even further, showing a man emerging from the cocoon of a gorilla skin. Rejected for exhibition at the American Academy of Design in 1924, *The Chrysalis* stirred up public debate about evolution at a time when religious fundamentalists in the U.S. were poised to orchestrate the notorious Scopes 'Monkey' trial. The Reverend Francis Potter of New York's West Side Unitarian Church, a pro-Darwinian pastor, now made the extraordinary gesture of exhibiting Akeley's controversial sculpture in his church and inviting Akeley to 'address the Congregation on "Personality in Animals," and . . . defend gorillas and other animals from the charges of "bestiality" which . . . are frequently made by opponents of evolution'.[43]

Taking his reappraisal of the gorilla a step further, Akeley returned to Africa in 1926, accompanied by a team of taxidermists and artists, with the aim of documenting every aspect of the gorilla's environment in minute detail. After Akeley sickened and died on this trip, his wife, the naturalist Mary Jobe Akeley, took over the expedition and completed Akeley's vision of recording the exact locale in the Virungas on the slope of Mt Karisimbi where he had shot his first male gorilla in 1921. She supervised the collecting of 22 species of plant from this location and the production of hundreds of plaster casts of their leaves and stems. These, combined with artist William R. Leigh's panoramic backdrop, based upon studies painted at Mt Karisimbi, provided the most naturalistic display of the gorilla ever mounted in a natural history museum.[44] The mountain gorilla diorama in the American

Great Apes in the Mikeno-Karisimbi Forest, c. 1927, diorama, American Museum of Natural History, New York. Gorillas shot by Carl Akeley, diorama painted by William R. Leigh.

Museum of Natural History's Akeley Hall of African Mammals remains one of that institution's signature displays. As Henry Fairfield Osborn noted in his introduction to *In Brightest Africa*:

> Akeley's work . . . defends the reputation of this animal, which has been misrepresented in narrative and fiction as a ferocious biped that attacks man at every opportunity, abducts native women as in the sculptures of Fremiet, a monster with all the vices of man and none of the virtues. For this untruthful picture Akeley substitutes a real gorilla, chiefly a quadruped in locomotion, not seeking combat with man, ferocious only when his family rights are invaded, benign rather than malignant in countenance. Thus he explodes the age-long gorilla myth and we learn for the first time the place in nature of this great anthropoid and come to believe that it should be conserved and protected rather than eliminated.[45]

Carl Akeley returned from the 1921 expedition to Africa convinced that 'the killing of a reasonable number of [gorilla] specimens for scientific institutions is legitimate and necessary, but the indiscriminate killing by sportsmen and others is unpardonable'.[46] His experience of filming the mountain gorilla in the wild provided him with the argument that this was the only way to 'shoot' gorillas from then on. He was subsequently to successfully lobby the Belgian government and Belgium's King Albert for the foundation in the Belgian Congo of Africa's first national park. Akeley first announced his vision for the sanctuary, 'a place where scientists may live and study gorillas', in an address to New York's Academy of Natural Sciences in December 1922; and in 1925 a 60,000-acre (250-sq. mile) reserve spanning what is today Rwanda, Uganda and the Democratic Republic of the Congo was set aside for this purpose. This was expanded to 400,000 acres in 1929.

A retired real estate man from Jacksonville, Florida, who turned big game hunter and stalked prey across the globe, Ben Burbridge visited Africa repeatedly from 1922. It was on two of these trips that he captured eight live gorillas, a record for the time – three of which survived to be successfully relocated to Europe and the U.S. Burbridge set out to capture gorillas on film as well, and initially intended to shoot them only this way.

'The Author and a Captive Gorilla', from Ben Burbridge, *Gorilla: Tracking and Capturing the Ape-man of Africa* (1928).

A man of his times, however, he was conditioned to view them as monstrously aggressive; he later told a reporter how

> One of the things he learned is that the natives who told him that gorillas would attack human beings were right. 'I didn't intend to kill any gorillas,' Mr Burbridge told me, 'but I did kill three, each time in self-defence as the beasts charged me.'[47]

In viewing the gorilla's bluff charge as a genuine physical threat, Burbridge reflected the still-dominant view of the animals; later he would express regret for these shootings, and recommend that the Belgian government increase the size of its gorilla sanctuary.

Burbridge developed a technique of panicking gorilla groups by mimicking the calls of their only non-human predator, the leopard; in the ensuing melee infants became separated from their parents and could be sometimes be snatched away. On his first gorilla expedition, in 1922, all but one of his captives died on the long march from the Virungas through British East Africa to the Indian Ocean. The surviving gorilla was sent to the Antwerp Zoo, where it died a year later. His expedition of 1925 yielded two more live exports – one again sent to Antwerp where it quickly died, and a young female which he shipped back to the States, placing her in the care of his brother James and sister-in-law Juanita in Jacksonville. Christened Miss Congo, this 'gorilla flapper' delighted her carers with her human-like traits – deep affection, curiosity, exhibitionism and jealousy of any attention paid to rival animals, such as the family dog. She also demonstrated her cognitive intelligence and dexterity, learning quickly to slide the catch on her restraining collar. A born showman, Burbridge promoted Miss Congo as being the only living gorilla to be seen outside of the African forests.

News of Miss Congo, the first female mountain gorilla (*Gorilla beringei*) to be transported to America, brought psychobiologist Dr Robert M. Yerkes down from Yale University to examine her relative intelligence. Study of the behaviour of the mountain gorilla then being in its infancy, having been only recently begun during Carl Akeley's African expedition of 1921, Miss Congo's presence offered scientific researchers unprecedented opportunities. Yerkes spent six weeks in Jacksonville putting the young enforced immigrant through her paces; his observations would be published in 1927 as *The Mind of a Gorilla*, the first study of its type. He had originally published a stacked-boxes and suspended-fruit intelligence test in his monograph *The Mental Life of Monkeys and Apes* of 1916, using monkeys and an orang-utan as subjects. Now he watched Miss Congo sequentially ladder-stack three randomly placed boxes to reach an orange hung high in the air, or use a pole to prise out an apple placed deep within a pipe. Other tests involved memory, in which she successfully dug up a favourite food that she had observed being buried several days previously; and recognition, when she was confronted with the conundrum of her reflection in a mirror. While Yerkes bonded affectionately with the young gorilla, he nonetheless found her intelligence disappointingly lacklustre compared to his earlier findings with chimpanzees. A strongly hierarchical thinker directed by ideas of ranking intelligence (an approach that would see his reputation sullied later by implications of racist eugenics), Yerkes ultimately found Miss Congo wanting.[48]

As an avid hunter who had shot gorillas both mortally and cinematically, Ben Burbridge followed up his experiences in the Congo with film and book deals. In December 1926 his commercially released silent feature *Gorilla Hunt*, which combined the big game and expeditionary genres, documented his hunting expeditions in British East Africa as well as the subsequent adventures of Miss Congo 'back home'. The *New York Times* review underlined the gorilla's playfulness and intelligence:

> Mr Burbridge shows some amusing scenes with these animals, one of them being that of a young gorilla who insists on getting tangled up in a drum of film . . . An extraordinary scene illustrates one captive gorilla trying to reach the branches of a tree. He is evidently thinking. He has a small box, which is not high enough for his intent, and as soon as he espies another box, he puts the two boxes on top of each other and his perspicacity is rewarded.[49]

After serializing a full account of his adventures in the journal *Forest and Stream* in 1926–7, Burbridge published them in book form in 1928 as *Gorilla: Tracking and Capturing the Ape-man of Africa*. He especially liked to recount for journalists his more hair-raising experiences with the great apes, such as how 'my twisted thumb and broken knuckles will bear the scars of battle with the young gorilla Bula Matadi [Rock Crusher] that I finally succeeded in capturing . . . the largest gorilla ever known to have been taken alive'.[50]

Under the care of the Burbridges, Miss Congo quickly grew from 20.4 kg (45 lb) on her arrival in Jacksonville to a weight of more than 45.4 kg (100 lb). When she inevitably became too large and strong for them to handle, she was transferred to the Ringling Bros. and Barnum & Bailey Circus, which had winter quarters in Sarasota, Florida, where she was to be groomed as a future act for the company. Here she was visited again by Yerkes for more intelligence testing. Miss Congo went into decline after being moved from Juanita Burbridge's care to the circus, and died on 24 April 1928. While Yerkes himself noted how Miss Congo seemed to be 'filled with infinite loneliness', *Time* magazine refused

Poster for *Congorilla* (1932).

to allow that any human-like emotions may have contributed to her death. For *Time*, 'every death has its legitimate biological cause; homesickness and heartache are not included'; and for Miss Congo 'the immediate cause of death was colitis'.[51]

Martin and Osa Johnson, a husband and wife filmmaking team from Kansas, also took remarkable early footage of gorillas in the wild. The Johnsons met Carl Akeley at the New York Explorer's Club in 1921 when he had just returned from his first gorilla-focused expedition to the Congo. Akeley, dreaming of completing the dioramas for the American Museum of Natural History's African Hall, saw that he could harness the Johnsons' popular and commercial appeal to generate interest and financial support for the project. A strange alliance ensued, with patrons of the Museum sending the Johnsons to Africa and lending them credibility, while the filmmakers fabricated narrative and staged incidents where necessary to create suspense and ensure a mass market. After some hiccups, *Simba* was the result and a huge commercial success; released in 1928, it took $2 million at the box office. This led the Fox Film Corporation to agree to finance the Johnsons' first sound film in Africa, *Congorilla* (1932), featuring the gorillas and pygmies of the Ituri Forest in what was then the Belgian Congo. With Hollywood studio backing, the Johnsons were free to pursue entertainment above education, with sequences such as that where Osa plays jazz music on a phonograph to bemused Mbuti pygmies. Although they had trouble capturing gorilla vocalizations with their cumbersome sound equipment, the Johnsons managed to film unique aspects of gorilla behaviour in the wild, including images of chest-beating and territoriality, as well as taking more family-oriented footage of captured gorilla infants being tucked into bed at their base camp.

Like Burbridge, the Johnsons came away from the Virungas convinced that some 2,000 mountain gorillas inhabited this region, a vast increase on Akeley's estimate that barely 100 survived. 'I know now,' Martin declared, 'after finding the large numbers which we did on our last safari and in nine different districts, that there is no chance of them becoming extinct, no matter how many may be shot.' They themselves were opposed to the shooting of any gorillas. While the Johnsons had taken rifles with them for self-defence, arriving with an impression of the gorilla as a 'fearsome looking beast' with a 'cold, cruel and murderous' face and 'something about those eyes that suggested an evil spirit', they departed Africa with a very different view. Having emerged unscathed from gorilla charges in which they had expected to be torn apart limb from limb, and finding no evidence of verified gorilla attacks upon humans, they became convinced

No Mail, No Love, No Nothin', postcard, USA, 1930s.

Gorilla stamp, Belgian Congo, 1959.

that 'the gorilla is not a dangerous animal. We certainly provoked them enough to have aroused the desire to kill, but we were not injured and never found it necessary to shoot to protect ourselves.' The Johnsons happily returned to the U.S. with two infants and a baby of the species, which ended up in zoos in San Diego and Washington, DC. Martin Johnson later professed to be 'truly sorry that I did take them away from their mountain side', writing that to see them 'imprisoned behind bars and steel netting in a place far too small, when only a few months ago they had the wide space of the Congo in which to roam, makes me regret their capture'.[52]

It was not until German-American mammalogist George Schaller's pioneering fieldwork from 1959 that a scientist lived extensively with gorillas, observing their social interactions, intelligence and behaviour in the wild. Schaller's *The Mountain Gorilla: Ecology and Behavior*, published in 1963, brought home to an international public for the first time that gorillas in the wild are gentle, intelligent, even compassionate beings.[53]

Trained in behavioural science, Schaller commenced his fieldwork in the Belgian colony of the Congo in 1959 at the age of 26; the following year, this became the independent nation of Zaire. He worked in the Virunga Volcanoes of Parc National Albert, established in 1925 by the Belgian colonial authorities as Africa's first national park (and the second in the world after Yellowstone National Park in the U.S.) to protect mountain gorilla populations, as well as other rare species and the area's extraordinary biodiversity. This became the Parc National des Virunga in 1960 in the newly independent Zaire, now the Democratic Republic of the Congo or DRC, and Parc National des Volcans in Rwanda when it in turn gained independence in 1962. Many local people, as well as national politicians, have seen the national park model of protecting wild nature to the exclusion of other considerations as a colonial legacy; historically there have been many incursions for cattle grazing and poaching.

From August 1959 to September 1960 Schaller observed gorillas intensively, producing the world's first 'detailed account of the ecology and behaviour of the mountain gorilla in its natural environment'.[54] Schaller had 314 direct encounters with eleven different groups of mountain gorillas, during which he clocked up 466 hours of close observation. Over time he gained an extraordinary degree of acceptance from the animals, which he attributed to his decision to adopt submissive and non-invasive behaviour; above all, he noted, 'I carried no firearms which might imbue my actions with [even] unconscious aggressiveness.'[55]

Schaller had been preceded in the field study of the gorilla by his countryman Harold C. Bingham, who together with his wife travelled to Parc National Albert in 1929 as part of a joint expedition of Yale University and the Carnegie Institute to record the social customs and behaviour of *Gorilla beringei* in nature by tracking its nesting patterns, excreta, and munchings and tramplings of jungle vegetation (these remnants of its passing being often as close as they came to the ape itself).[56] Earlier still, a remarkable appraisal of the gorilla had been attempted by the American adventurer Richard L. Garner in the 1890s. Installing himself in the Congo in a bamboo-roofed steel wire cage (which he named Fort Gorilla) for 112 days and nights in 1893, Garner eventually observed over twenty gorillas passing by his hideout. This, along with his observations of an infant gorilla that he held captive for a short period before its death (probably from incorrect feeding), as well as dialogue with local peoples, led Garner to dramatically revise then-current notions of the species. As he told the *New York Times* in 1894: 'The gorilla is described as being an exceedingly vicious and dangerous animal. I did not find him to be very vicious. None of them, at any rate, tried to attack me.'[57]

In contrast to these well-meaning but somewhat circumscribed efforts, Schaller's micro-documentation of the mountain gorilla at the end of the 1950s was astonishing, ranging from the frequency, weight, section count and position excreted of the creature's faeces to its patterns of sneezing, yawning, burping, wind-breaking, self-grooming and nose- and teeth-picking. He also analysed the gorilla's facial expressions, vocalizations and possible related emotions, as well as itemizing the nine individual acts which comprise the full gorilla chest-beating sequence, making it seem the animal kingdom's ritualized equivalent of the Maori haka – a display designed both to release tension and to intimidate strangers. Schaller undertook a full study of the gorilla's social behaviour, defining eloquently the leadership role of the silverback and decoding the sub-species' complex modes of dominance, mutual grooming, play, mating behaviour and responses to death of kin. Most significantly for Dian Fossey's later interactions with the mountain gorilla, Schaller decoded the species' various signs of both aggressive and submissive behaviour, writing of how while 'an unwavering stare is a form of threat . . . gorillas in the wild and in captivity frequently use head-turning to indicate submissiveness in response to a threatening stare by another gorilla or human being'.[58]

Noting how individual gorillas can be identified by their nose shape and markings as clearly as people, Schaller gave names to the animals encountered in his study groups, admittedly anthropomorphizing but also individualizing the gorillas for the first time – in a manner compellingly different from the random naming of captive infants then operating in circuses and zoos, which frequently served to mask knowledge of the trauma of their seizure and the slaughter of their parents. Above all, Schaller felt that he was witnessing at the eleventh hour a gorilla in crisis. Only 400 to 500 gorillas inhabited the Virunga Volcanoes, he argued, out of a sum total which he estimated at that time to be of the order of 5,000 and certainly no more than 15,000 mountain gorillas overall.

In 1964 Schaller summarized his findings in a popular eco-travel narrative, *The Year of the Gorilla*. In this he brought the thrill of encountering gorillas in the wild alive in a manner not permissible in his scientific writings:

> Accustomed to the drab gorillas in zoos, with their pelage lustreless and scuffed by the cement floors of their cages, I was little prepared for the beauty of the beasts before me. Their hair was not merely black, but a shining blue-black, and their black faces shone as if polished . . . I felt a desire to communicate with him, to let him know by some small

Mountain gorillas, Pablo Group, 7 February 2008.

> gesture that I intended no harm, that I wished only to be near him. Never before had I had this feeling on meeting an animal.[59]

He also argued passionately against the collecting of gorillas for zoos, revealing for the American public how:

> For each of the eighty-five gorillas alive in the United States today, at least five others died while being captured or before they reached a zoo, a sad commentary not only on many collectors, but also on zoos, which for the most part care little how their animals are obtained . . . one reliable authority told me that in about 1948 officials organized

> the killing of some sixty mountain gorillas near Angumu to obtain eleven infants for zoos.

Schaller was appalled by the fact that

> No zoo bred gorillas before 1956, a sad record considering the large number of these apes that over the years have been held in captivity. The treatment of apes in zoos has been and often still is scandalous. The creatures sit alone behind bars, like prisoners in solitary confinement.[60]

The gruesome facts of the gorilla trade for zoos and collectors had been in the public arena for some time. In 1941 *Life* magazine had reported on the activities of animal collector Phillip Carroll, who single-handedly doubled the population of captive gorillas in the U.S., importing eight infants from Free French Africa destined for zoos in St Louis, San Diego and the Bronx.[61] A decade later it followed another animal collector, Bill Said, on a gorilla hunt in French Equatorial Africa. Noting how 'Said is often forced to kill all the adult gorillas on the chance of capturing one or two marketable young ones', *Life* documented his slaughter of six gorillas in order to secure a pair of infants for American zoos. It also revealed that on two earlier hunts Said had captured six infants in an equally violent fashion (three of which were sold to the zoo in Columbus, Ohio, for $10,000).[62] The story and its brutal photographs prompted one reader to protest: 'One of the cruellest articles I've ever read . . . Come now, who in America wants to see a gorilla that badly?'[63] In John Ford's film *Mogambo* (1953), the big game hunter Victor Marswell (played by Clark Gable) shoots the leader of a gorilla pack as it charges at him while he is trying to capture young gorillas to sell. He drowns his sorrows that night, depressed at having wantonly killed the magnificent beast.

Gorilla Family, postcard, Spain, 1950s.

Schaller's documentation of the distinct personalities of gorillas and the close family bonds uniting members of their groups now brought home the immense cruelty of this animal trade, revealing its shattering of primate communities to be a form of mass murder. The world's conscience started to awaken. While the numbers of gorillas killed and exported was small compared to the hundreds of thousands of monkeys imported into the U.S. each year for medical research, readers were moved by Schaller's descriptions of the strong social ties uniting female gorillas and their offspring for years after birth, and the species' contentment with life in family groups under the protective custody of silverbacks whose gentler side was now made clearly known. His conclusion was as startling as it was simple: 'Gorillas are eminently gentle and amiable creatures, and the dictum of peaceful coexistence is their way of life. In this man would do well to learn from the gorilla.'[64] Schaller reflected, 'If nothing else, man should show some ethical and moral responsibility towards creatures who resemble him so closely in body and mind. But then man has never learned to treat even his own kind with compassion.'[65]

Dian Fossey's work built on George Schaller's groundbreaking fieldwork and brought the study of gorillas to a much wider public as *National Geographic* magazine articles, photographs and films made her a household name. Her gorilla-mimicking behaviour in the field to gain the confidence of the animals became legendary, and her violent murder in 1985 was front-page news.

An occupational therapist who had been encouraged to study the great apes while visiting Dr Louis Leakey in Tanzania in 1963, Dian Fossey commenced work in 1967 in the Virunga chain of volcanic mountains, initially in the Kabara meadow in what was then Zaire, the area where both Carl Akeley and Schaller had preceded her (today the DRC). However, in July 1967 she fled an army rebellion in Zaire's Kivu Province and moved to Rwanda. Here she established the Karisoke Research Centre in a saddle region between Mt Karisimbi and Mt Visoke, two volcanoes in the Parc des Volcans, where she continued to work for almost twenty years. Initially surveying mountain gorilla population numbers by counting the nests that they build from leaves and branches to sleep upon on the ground, she went on to develop her own observation methods and to provide revolutionary new documentation of gorilla behaviour. Making an intensive study of gorilla vocalizations, she used her imitations of this 'language' of sounds to gain an unprecedented degree of acceptance within gorilla groups.

Eschewing the 'sit and observe' approach previously established in primate fieldwork, Fossey 'tried to elicit their confidence and curiosity by acting like a gorilla'.[66] Lamenting that 'the gorilla is one of the most maligned animals in the world', Fossey challenged former notions of the 'white-fanged hairy ape-man' and championed a new vision of the 'introverted, peaceful vegetarian'.[67] Her landmark story, 'Making Friends with Mountain Gorillas', appeared in *National Geographic* magazine in January 1970. Here she wrote compellingly of her fears for the future of her 'forest friends', from whom she argued, 'after more than 2,000 hours of direct observation, I can account for less than five minutes of what might be called "aggressive" behaviour'.[68] Robert Campbell's enchanting cover photograph of Fossey walking in the Rwandan forest with a baby gorilla nestled in her arms helped catapult her to a worldwide celebrity that was to prove problematic.

Fossey took Schaller's practice of anthropomorphically naming specific gorillas within groups or 'families' a step further, naming every individual within a group and constructing generational family trees, proving 'how the strong bonds of kinship

Gorilla stamps, Rwanda, 1970.

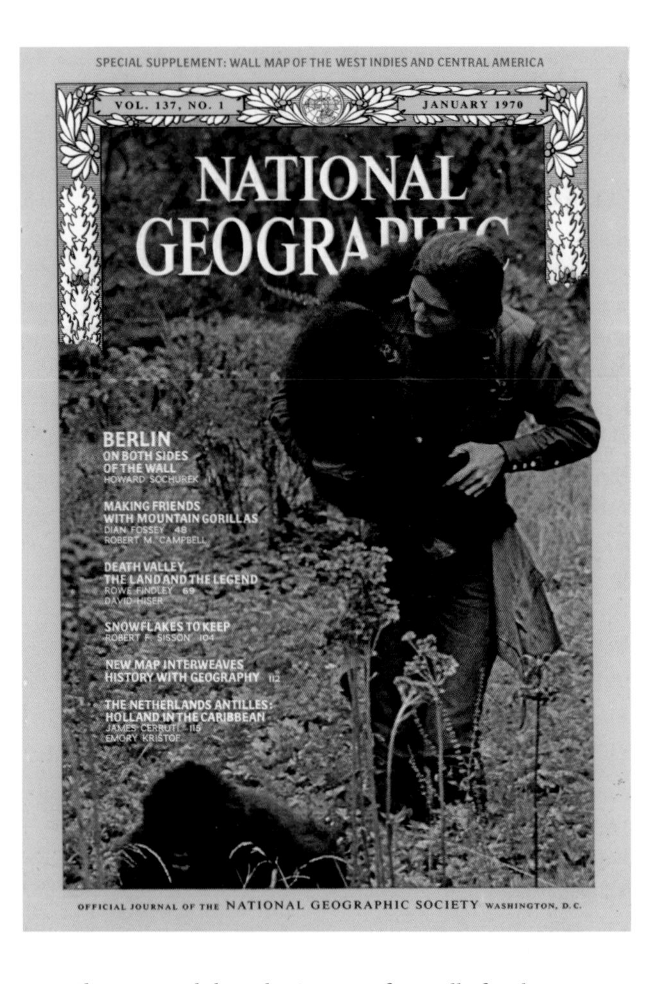

'Two Baby Gorillas Share a Forest Stroll with Author Dian Fossey', cover of *National Geographic* (January 1970).

contribute toward the cohesiveness of a gorilla family unit over time' and thereby humanizing the gorilla for new audiences.[69] Her first fifteen years of gorilla observations, which she summarized in her autobiographical account *Gorillas in the Mist* in 1983, advanced knowledge of the species and its behaviour enormously.

Whereas Schaller had seen no evidence of coprophagy during his time with the mountain gorillas, Fossey noted that 'all age and classes of gorillas have been observed eating their own dung and, to a lesser extent, that of other gorillas', an act that allows them to absorb nutrients not released in the first processing of plant matter.[70] Unlike Schaller, Fossey witnessed gorilla masturbation, as well as homosexual behaviour among both sexes. Schaller did not see any serious acts of aggression between gorillas within a specific group, and also felt that 'inter-group antagonism was rare, and was confined to threatening stares and bluff charges'.[71] Fossey, however, recorded numerous instances of inter-group violence. She also found infanticide to be a common occurence when a silverback paired with a new female who had infants in tow – a means of eliminating another male's line and ensuring the primacy of his own blood stock, which was accompanied by swift impregnation of the new female. Knowledge of these acts of intraspecies violence only served, ironically, to make the gorillas seem even closer to humanity in their behaviour.

As news of Fossey's discoveries and close encounters with habituated gorillas spread around the globe, it brought an unwanted influx of reporters, tourists and nature photographers to Karisoke. Her decision in 1972 to allow an image of a favourite young gorilla, Digit (named on account of his having an old finger injury), to be used on a poster advertising the Parc des Volcans issued by the Rwandan Office of Tourism soon led to an ominous 'feeling that our privacy was on the verge of being invaded'.[72] She blamed harrassment from a French film team for a miscarriage experienced by Effie from Group 5, and also felt that relentless new incursions by strangers into the gorillas' mountain habitat increased the risk of human diseases being transmitted to 'her' apes.[73] Her own students were chastised severely for defecating near gorilla groups.

Dian Fossey with Digit (who was killed in 1977).

Digit's violent killing by poachers (decapitated, with his hands cut off and his body riddled with spear wounds) on 31 December 1977 shattered Fossey, who now thought that her trusted proximity to the mountain gorillas 'was a privilege that I no longer deserved'.[74] Renowned American newscaster Walter Kronkite announced Digit's death on CBS television two months later, on 3 February 1978, to millions of viewers who had 'befriended' this gorilla through *National Geographic* broadcasts.[75] The slaughter and decapitation in July 1978 of Uncle Bert, Digit's 'uncle' from Group 4, was another devastating blow. In the wake of Digit's death, Fossey organized the Digit Fund, which aimed to promote active conservation through the systematic elimination of poaching. She also used the media attention she now commanded to this end. 'It was Digit, and he was gone. The mutilated body, head and hands hacked off for grisly trophies, lay limp in the brush like a bloody sack.' These opening words of Fossey's *National Geographic* article 'The Imperilled Mountain Gorilla' of 1981, with its photographic documentation of named gorilla individuals slaughtered by poachers, sounded an alert about the massive threat posed to the mountain gorilla not only by local poachers, but by other forms of 'human encroachment, land clearing, illicit collecting' and relentless tourism.[76] In *Gorillas in the Mist*, Fossey was to shock the world with accounts of poachers culling gorilla ears, tongues and testicles for brewing into a potion that she said African *sumu* black magic believed endowed its recipient with the prowess and virility of a silverback. Unwanted gorilla hands could be sold to tourists for a mere $20. However, she made a point of stressing that both Digit and Uncle Bert had died defending their families from poachers who were probably attempting to kidnap infants, rather than having been singled out for the trophy hunting of their heads. Fossey had had previous experience of the hunting of gorillas for zoos, recording how around eighteen gorillas from two groups had been slaughtered in order to obtain two infants for the Cologne Zoo. Fossey herself cared for these traumatized gorilla babies while vainly attempting to thwart their export, and later learned that they both died within a month of each other in Cologne in 1978. Because of these experiences, Fossey controversially refused to habituate gorillas to her black African staff, arguing that 'the second that it takes a gorilla to determine whether an African is friend or foe is the second that might cost the animal its life from a spear, arrow, or bullet.'[77]

Fossey's *Gorillas in the Mist* was to become the best-selling book to date on the gorilla, and remains so. Fossey's murder at Karisoke in 1985 fuelled debate about both poaching and the negative effects of tourism on dwindling gorilla populations (while the first accounts attributed her death to vengeful poachers who resented her protection of the gorilla groups from them, theories were later to multiply linking her death to those who wanted greater tourist access to the gorilla for commercial gain). Dian Fossey's cinematic beatification by actress Sigourney Weaver in the feature film *Gorillas in the Mist* (1988) marked a zenith of concern among a broad international public for the plight of the threatened gorilla. Here, interestingly, *National Geographic* footage of real gorillas was interspersed with sequences in which actors were used to play the adult gorillas, including Digit interacting with other actors in the film. The orphan gorillas destined for Cologne were played by young chimpanzees disguised with black make-up and fur.

The legacy of Fossey's researches has been kept alive, apart from the perennial popularity of her autobiographical writings, by documentary films devoted to her original study of gorillas and their descendants in the Virunga mountains. In 2006 Sigourney Weaver returned to Rwanda to film the BBC Natural History and Animal Planet documentary *Gorillas Revisited*, which retold the Dian Fossey story for a new generation (not be confused with the BBC's *Gorillas Revisited with David Attenborough* of 2005). The PBS film *Titus: The Gorilla King* (2008) told of the challenge of three decades of survival faced by the silverback Titus – son of the Uncle Bert slain by poachers in 1978 – up until his own displacement by a younger silverback rival. This story in turn was updated in the BBC's *Mountain Gorilla* trilogy of 2010, narrated by Patrick Stewart.

Subsequent to Schaller's and Fossey's groundbreaking researches, naturalists and field observers have increased and refined our knowledge of gorilla behaviour. Expanding upon Fossey's investigations into gorilla vocalizations, Alexander Harcourt and Kelly Stewart have further decoded the apes' close-calls or low throat noises made when in close proximity with one another as performing specific communicative functions, such as 'a sort of collective vote on when to end a resting or feeding period'.[78] And whereas Schaller noted how mountain gorillas manipulated prickly plants, folding their stinging sections inwards when eating to avoid mouth injury, Richard Byrne has subsequently itemized a complex host of separate leaf manipulations used by these gorillas.[79]

While Fossey was vehemently opposed to mountain gorilla tourism, today she is cited repeatedly on the Rwandan tourist board's website. Tourism is now generally promoted by conservationists and government officials alike as a necessary revenue stream that also raises awareness to help save the critically endangered sub-species. The remaining populations are found in the area of the Virunga mountains – ranging across Rwanda, Democratic Republic of the Congo and Uganda – and in Uganda's Bwindi Impenetrable Forest, a short distance to the north. Mountain gorillas can only be seen in the wild, as they do not survive in zoos.

The Rwandan government issues permits to visitors wishing to trek in the Parc des Volcans and spend an hour with the gorillas in their bamboo forest habitats. During the Rwandan civil war the park was occupied by military forces and all gorilla research and tourism stopped for around five years, recommencing in 1999. In 2010 there were seven to eight family groups of gorillas in the park varying in size from around 10 to 40 gorillas; each group could be visited by up to eight tourists per day. Individual gorillas and the families are named, as previously by Fossey, and in 2005 the Rwandan government established the annual Kwita Izina gorilla baby naming ceremony.[80] In Uganda, the Mgahinga Gorilla National Park, covering the three northernmost Virunga Volcanoes, and the Bwindi Impenetrable National Park also offer gorilla tracking and issue permits to visit one of the habituated family groups with a guide.

Across the border in the Democratic Republic of the Congo's Virunga National Park, home to both mountain gorillas and Eastern lowland gorillas, a highly unstable political situation has had a severe impact. Beyond the major threats posed by the illegal trade in bushmeat and live animals, habitat destruction and illegal exploitation of minerals (notably Coltan, used in mobile phones, and gold), more than a decade of civil war over the turn of the millennium and later military activity and civil unrest have directly affected gorilla populations. After the genocide in Rwanda in 1994, Hutu militiamen and refugees fled across the border. The park was then used as a hideout and battleground for rebel groups, including the Democratic Forces for the Liberation of Rwanda, former Congolese president Laurent Kabila's armed forces, which overthrew dictator Mobutu Sese Seko, and, after he broke away from Kabila, rebel Laurent Nkunda's National Congress for the Defence of the People. Park rangers worked through the civil war without receiving a salary for long periods, and over 160 rangers were killed in the area in the first decade of the new millennium.

Large numbers of refugees displaced by the fighting settled on the southern border of the Virunga National Park around the regional capital of Goma with no access to fuel for cooking other than buying sacks of charcoal made from the park's forests. The charcoal trade is illegal and much of it is controlled by armed rebels. Rangers are attempting to stop the trade by destroying production kilns in the forest. It is believed to be in retaliation for the efforts of one such ranger, Paulin Ngobobo, to stop charcoal production that a series of gorilla murders took place on 8 June and 22 July 2007: six were shot, including a silverback male and a pregnant female, their bodies left as a warning to rangers. News of these events shocked the world;[81] a subsequent National Geographic Television documentary, *Explorer: Gorilla Murders*, was awarded an Emmy for Oustanding Investigative Journalism. Less than a year later, in 2008, a new Chief Warden of the park, Belgian conservationist Emmanuel de Merode, was appointed by the governing body, the Congolese Institute for the Conservation

A young Western lowland gorilla.

of Nature (ICCN). Since then a range of initiatives have been put in place, including a website, blog and iPhone App to promote the park and gorilla conservation, allowing an international public to follow the lives of the gorilla families and the activities of the rangers. The park management is also trying to address the needs of local populations and offer alternatives, for example helping people to switch from charcoal to fuel briquettes made from sawdust, rice husks, leaf mulch and other organic waste.

Mountain gorillas have had a very high profile in popular reporting on the gorilla because of the involvement of figures such as Fossey as well as, in an earlier period, Carl Akeley, with his celebrated mountain gorilla diorama for the American Museum of Natural History in New York. Important gorilla conservation sites exist across all of the major areas of gorilla habitation and today these share most of the challenges faced by Africa's oldest national parks in the Virunga area, home to the mountain gorillas. The United Nations declared 2009 the Year of the Gorilla to raise awareness and help park authorities in the gorilla-range countries to advance its conservation status. Western lowland gorillas face the particular problem of higher exposure to the Ebola epidemics devastating great ape populations. Other human diseases to which gorillas are susceptible include respiratory viruses and gastrointestinal pathogens; gorilla tourism can run the risk of exposing them to these. Throughout the remaining gorilla habitats, deforestation is also a major problem for all sub-species, as is the illegal wildlife and bushmeat trade.

3 Gorilla Mania

And there came a moment a few years ago when an American traveller, Mr du Chaillu, returning from the western part of equatorial Africa, brought the remains of many gorillas to London, and recounted marvellous things about this horrendous ape to those hungry for titillating news. The gorilla became the beast *à la mode*, the favourite of the newspapers and of fashionable city talk, constantly appearing in its place of honour between discussions of cotton and anti-French politics.
Filippo De Filippi, 'L'Uomo e le scimie' ('Man and the Apes'), public lecture delivered in Turin, 11 January 1864[1]

From nineteenth- and early twentieth-century stories of gorilla encounters and publications by European and American writers and scientists there emerged a 'gorilla mania' in art, literature and popular entertainment which quickly lost any regard for realism or authenticity. This fascination began as early as the international 'unveiling' of the gorilla itself, when a voracious public in the U.S., London and Paris avidly consumed the writings and lectures of Paul du Chaillu, enthralled by his theatrical accounts of gorillas chomping on rifles like carrots. Feted in London as 1861's 'Lion of the Season', Du Chaillu became the topic of songs and musical entertainments.[2]

Between 1867 and 1872, Paul du Chaillu recast his explorations in Gabon as narratives for younger audiences. Capitalizing on his fame as the first European to describe gorillas in their native habitat, he wrote the first of his children's books, *Stories from the Gorilla Country* (1867), to convey 'the vigorous quality of romantic exploration and the color and dramatic interest of wild life in the jungle' – it was a huge commercial success.[3] Following it in rapid succession with *Wildlife Under the Equator* (1868), *Lost in the Jungle* (1869), *My Apingi Kingdom* (1870) and *The Country of the Dwarfs* (1871–2),

Alfred Concanen and Thomas Lee, *The Gorilla Quadrille*, lithograph, cover of sheet music by Charles Handel Rand Marriott, London, 1861.

John Leech, *The Lion of the Season: Mr G-g-g-o-o-o-rilla!* Cartoon satire of Paul du Chaillu, electrotype, published in *Punch* (25 May 1861).

the now middle-aged 'Gorilla Hunter' was adored by thousands of young readers.

Du Chaillu's writings were translated into French and, serialized in popular Parisian magazines like the *Journal des voyages*, they appealed equally to audiences old and young. They fitted well within the *Journal des voyages*, which splashed chest-thumping gorilla fiction across its pages for more than three decades following its founding in 1877. If a story merged fact and fiction, so much the better, thrilling and incredible adventures providing the lifeblood of this publication.

This whole genre of gorilla adventure tales, created by Paul du Chaillu and refined by subsequent story-tellers, also filtered down into the immortal Tarzan narratives of Edgar Rice Burroughs, which in the early twentieth century were to spawn a related publishing and cinema phenomenon full of rippling primate action.

It informs too the popular Don Sturdy series of children's books, published in the U.S. for a decade from the mid-1920s, which were ostensibly by Victor Appleton but actually ghostwritten by John W. Duffield. *Don Sturdy Among the Gorillas*, the seventh of these fifteen rollicking yarns, released in 1927 with illustrations by Walter S. Rogers, was one of the series' most successful titles. Meanwhile, in Great Britain and Australia, Robert Ballantyne's 1861 classic *The Gorilla Hunters* remained perennially in print throughout the 1920s and '30s, alongside the Du Chaillu tales.

Cover design for Paul du Chaillu, *Stories of the Gorilla Country, Narrated for Young People* (1867).

As the comic book format developed in the U.S. in the second half of the 1930s, these action-packed stories transferred directly across into their new abbreviated visual formats. A number of gorilla-themed titles in the comic's 'Golden Age' (1938–55) were inspired by the Hollywood film serials that were so popular at this time. Nyoka the Jungle Girl, a character originally introduced

Death of a Male Gorilla, wood engraving, from Paul du Chaillu, *Wildlife Under the Equator* (1868).

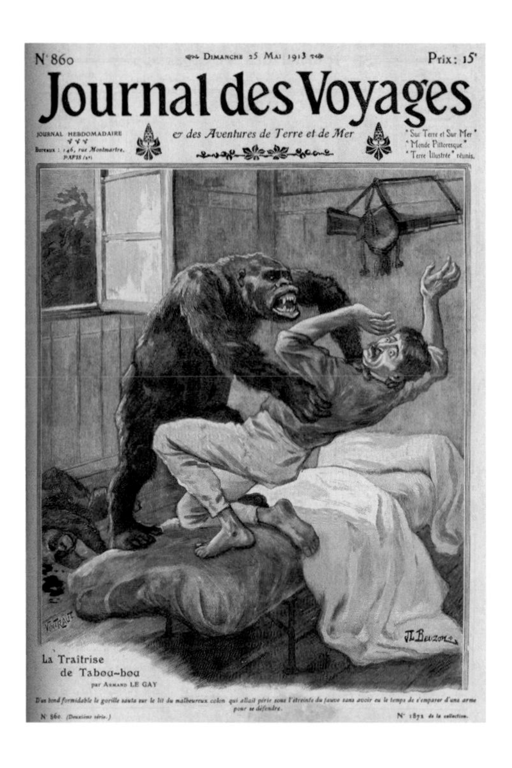
N°860 Dimanche 25 Mai 1913 Prix: 15c

Journal des Voyages

Journal hebdomadaire

Bureaux : 146, rue Montmartre, Paris (2e)

& des Aventures de Terre et de Mer

"Sur Terre et Sur Mer" "Monde Pittoresque" "Terre Illustrée" réunis.

La Traîtrise de Tabou-bou par Armand Le Gay

D'un bond formidable le gorille sauta sur le lit du malheureux colon qui allait périr sous l'étreinte du fauve sans avoir eu le temps de s'emparer d'une arme pour se défendre.

N° 860 (Deuxième série.) N° 1872 de la collection.

Cover of *Journal des voyages*, wood engraving (25 May 1913).

by Edgar Rice Burroughs in a short story of 1929, appeared in a number of titles published by Fawcett between 1944 and 1953. These were offshoots of two Republic Pictures film serials, *Jungle Girl* (1941) and *Perils of Nyoka* (1942), in each of which the heroine was menaced by a grinning gorilla (animal impersonator Emil van Horne in his distinctively comic furry suit). Nyoka had been preceded in the late 1930s by Sheena, Queen of the Jungle, a sort of female Tarzan; and she was followed by Princess Pantha in 1946, Rulah, Jungle Goddess, in 1947, and Tegra (later Zegra), Jungle Empress, in 1948. All these feisty jungle women saw significant gorilla action.[4]

Cover of *Nyoka the Jungle Girl* (September 1947).

In the 1950s gorillas appeared regularly on the covers of DC Comics's *Strange Adventures*, accompanying bizarre stories of evolutionary ray machines, alien gorilla races and mental transference between humans and apes. Mind-melding propelled the menacing gorilla action of Congo Bill (a DC Comics character who also featured in Columbia Pictures's fifteen-chapter serial from 1948) into new realms in *Action Comics* in the late 1950s, when a dying witch doctor friend gave the hero a magic ring. When rubbed, this exchanged his consciousness with that of a giant Golden Gorilla to create the mighty entity known as Congorilla. In a similar transmogrification, in 1954 the Marvel Comics character Ken Hale turned into the superhero Gorilla Man while hunting apes in Africa. Gorilla characters were inserted into other superhero comics of the 1950s, such as Archie Comics's *The Fly* and DC Comics's *Superman's Pal Jimmy Olsen*, to boost flagging sales.[5] One of these, the telepathically evil Gorilla Grodd, who first appeared in DC Comics's *The Flash* in May 1959, developed supervillain status, starring over subsequent decades in numerous other comic titles as well as in several television series and finally in video games (a medium in which gorilla-like apes were also to feature heavily in Nintendo's *Donkey Kong*).

One of comic-land's most charismatic if troubled heroes was Konga, who appeared as the eponymous hero quasi-synchronously in a British sci-fi film (released by Anglo-Amalgamated in 1961) and a Charlton Comics title (22 issues between June 1960 and November 1965). In the film Konga is a hapless chimp who is transformed into a monstrous giant after injections of a serum made from mutant African carnivorous plants. At first used as an assassin by his maker, mad botanist Charles Decker, Konga grows confused and rebellious as he shoots up in size to more than 30 metres, eventually killing everyone in his path and destroying large swathes of central London. Whereas the low-budget cinematic Konga resembled something salvaged from a hard rubbish drop, the comic-book hero was an ape whose appearance could be as noble as his innately gentle character. 'Fleeing from the civilization that wishes to kill him, from mankind who brings him pain and hurt', Konga seeks peace on an island near the equator or in the freezing cold of Antarctica while surviving troubles as varied as attacks from interplanetary aliens, prehistoric atomic mutants from the centre of the earth, the Cuban Missile Crisis and those dreaded Commies.[6]

Cover of *Konga*, Charlton Comics (March 1962).

The *Konga* comics offered advertisements for mail-order toy soldiers, how-to-draw kits featuring scantily clad female models,

Based On An American International Picture
STILL 10c
KONGA
A Charlton Publication
APPROVED BY THE COMICS CODE AUTHORITY
MARCH
from the Herman Cohen Production
THE PRIMEVAL BEAST FROM BEHIND THE VEIL OF TIME FACES MAN'S SUPER-WEAPON...
THE H BOMB
TALLER THAN THE HIGHEST BUILDING...
STRONGER THAN A HERD OF ELEPHANTS...
KONGA!

Cover of *Man's Life* (March 1958), image by Wil Hulsey.

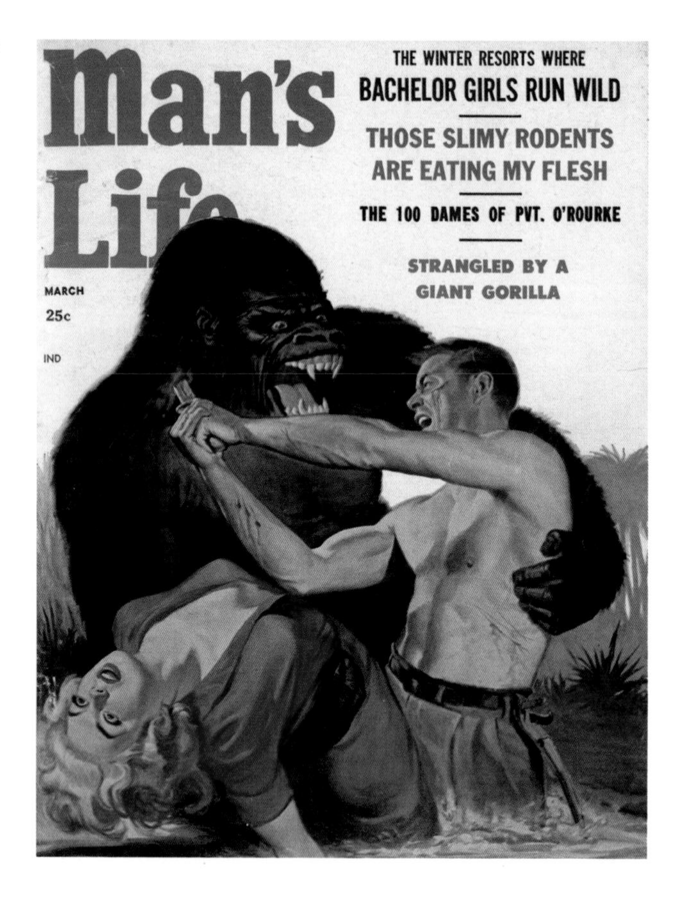

men's physique regimens and reproduction guns. Similar advertisements, minus the toys and with real guns, populated the adult male-oriented pulp adventure and thrill magazines that proliferated in the U.S. in the late 1950s. Seemingly also partially pitched at late teenagers (even if they had to sneak them from their fathers' dens), these frequently carried gorilla covers and stories, usually involving the abduction and heroic rescue of the man's woman.[7]

These were comics for grown-ups (or would-be grown-ups), guilty escapist pleasures targeting conventional nuclear-family husbands of the era and their post-pubescent sons. Today many of them, such as the issues of *True Men* and *Man's Life* with cover art by Will Hulsey, are considered 'animal attack masterpieces' by aficionados of the genre.[8]

Gorilla heroes and anti-heroes continue to be invented or re-invented for comics enthusiasts. In 1993 Art Adams created *Monkeyman*, the tale of a colossal super-intelligent gorilla zapped to earth from another dimension. *Congorilla* reappeared in 1992, while *Gorilla Man* was given a new lease of life in 2010 with a spectacular three-part Marvel Comics re-issue written by Jeff Parker and drawn and coloured by Giancarlo Caracuzzo and Jim Charalampidis. A gorilla hero, Captain Congo, was created by writer Ruth Starke and illustrator Greg Holfeld in Australia, with the first volume published in 2008, *Captain Congo and the*

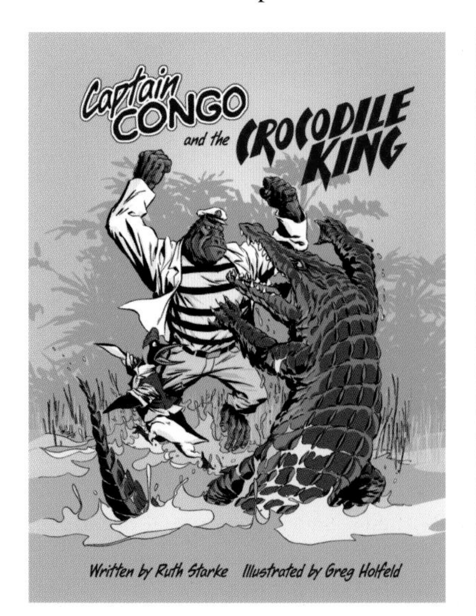

Cover of *Captain Congo and the Crocodile King* (2008), image by Greg Holfeld.

Cover of *Rage For Men* (March 1961).

Crocodile King. The gorilla hero Captain Congo here travels to Abyssinia with his penguin sidekick to find a missing anthropologist. In 2004 Aneurin Wright's *Lex Talionis* (Law of Retributive Justice) offered a starkly beautiful new perspective on the gorilla comic, blending the creator's own love of linoleum reduction prints with a jump-cut approach to narrative and a landscape format, and echoing classic French *bandes dessinées*, Japanese *manga* and Frank Miller's film noir ethos. Wright's dramatic narrative of a silverback delivering revenge upon a hunter who had killed and pillaged his family group near Rwanda's Mt Visoke delivers a sober and adult message inspired by the writings of George Schaller and Dian Fossey.[9]

To a large extent the world's fantasies about the gorilla's power and potential have been fuelled by its frustrated lack of access. In

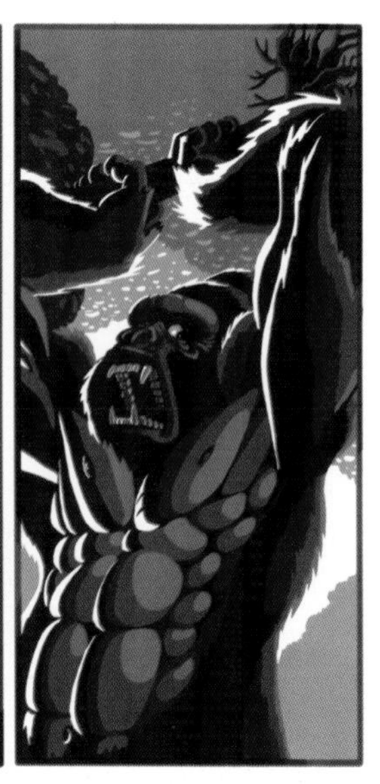

Aneurin Wright, *Lex Talionis: A Jungle Tale*, Image Comics (2004).

Gorillas Hall, Museum of the Belgian Congo, Tervuren, Belgium, postcard, *c.* 1900.

1901 the *Los Angeles Times* wrote of how 'the man who should succeed in capturing a fully-grown male gorilla alive and transporting him safely to civilization could command practically his own price'.[10] Inspired by the exotic stories of gorillas in the wild brought back by hunters and adventurers, the average city dweller could only contemplate the rather stiff taxidermied specimens that comprised most museums' gorilla displays. The world's desire to experience the thrill of gorillas at first-hand was to lead to a generations-long drive to bring live gorillas to their adoring public in zoos. In the realm of fiction this dream was played out on the big screen with the transportation to New York of a giant gorilla in the cinematic milestone *King Kong* of 1933.

Throughout the second half of the nineteenth century, most efforts to transport living specimens of the gorilla to Europe were doomed to failure. Writing in 1885, a specialist at the University of Berlin concluded that

> Every attempt to embark them immediately after their capture, without previously weaning them from their old modes of life, and adapting them slowly and systematically to their altered conditions, has invariably resulted, sooner or later, in sickness and death.[11]

While the gorilla had by then become a familiar object of both scientific and public discourse, living specimens of the great ape were still highly prized and exotic items for those zoos lucky enough to possess one. It was with genuine excitement that *The Times* informed its London readers in October 1887:

> For the first time since the establishment of the Gardens of the Zoological Society a living gorilla has been added to the collection . . . Although it has been scarcely a month in the gardens it is rapidly recovering from the shyness before strangers which it exhibited at first, and it feeds freely on almost every kind of fruit offered to it, showing a marked preference, however, for pomegranates.[12]

The zoo's director, Abraham Bartlett, wrote of how on arrival 'the poor beast appeared to be completely exhausted and almost lifeless – no doubt partly from exposure to the cold and the shaking noise of the railway journey.'[13] This infant, who was nicknamed

Mumbo, bonded with Bartlett but died after a few months (not surprising, perhaps, given his diet of cheese sandwiches, beer and sausages); his remains were sold to the Royal College of Surgeons and displayed in their museum.[14]

Another gorilla had docked in the UK slightly earlier than Mumbo, but only as a temporary visitor. This was Pongo, a young male brought back from West Africa by the German African Society's Expedition on the steamship *Loanda*, which stopped briefly in Liverpool in June 1876 before proceeding to Berlin. Found chained up in a village on the Gaboon River, brought to Germany by a Dr Falkenstein and sold to the Berlin Aquarium for 20,000 marks, this infant became the first publicly advertised living gorilla exhibit in Europe.[15] In twelve months more than 300,000 people visited Pongo the celebrity ape to be amazed by his human-like antics (which included, sadly, drinking alcohol and smoking). Pongo was subsequently loaned to a London pleasure palace, the Royal Aquarium in Westminster, for the summer of 1877, where he also attracted record crowds. *The Times* felt that 'as he has so early learnt both to smoke and drink, it is to be hoped that he may soon acquire the other accomplishments which distinguish civilization'.[16] Shortly after his return to Berlin, in November 1877, Pongo died. Preservation of his body post-mortem was not deemed essential, there being already 'plenty of stuffed skins and of skulls of gorillas in Germany, the result of Güssfeldt and Lenz's expeditions'.[17]

A baby female gorilla was exhibited at the Crystal Palace, along with an infant male chimpanzee companion, for a short period in the autumn of 1879, before succumbing to a lung infection. She had been imported by a naturalist in Liverpool and then sold on to a Manchester client for the fabulous sum of £2,000.[18] The first live gorilla to have been exhibited in Paris arrived a little over four years later, in April 1884. Another infant male from the Gabon, it was described as 'savage, morose, and brutal'.[19] It was to survive only a few days.

The first celebrity gorillas in the U.S. were in fact impostors. Being already possessed of an adult male chimpanzee, Chiko, the proprietors of the Barnum & Bailey Circus resolved to procure him a mate. In December 1893 a female primate, Johanna, was duly shipped from Lisbon to be his bride, and widely proclaimed to be a gorilla. Despite both animals obviously being chimpanzees, the circus insisted on advertising Johanna as a gorilla. Such deception was nothing new, P. T. Barnum having claimed to have added a live gorilla to his repertoire as early as 1867. No less an authority than Paul du Chaillu was called in to adjudicate on the status of Chiko and Johanna. While praising them as 'the largest

'The Third Gorilla Shipped to the UK (with Chimpanzee Companion) at the Crystal Palace', wood engraving, published in *The Graphic* (13 September 1879).

R. H. Moore, 'The Gorilla, "Pongo", at the Royal Aquarium, Westminster', wood engraving, published in *Illustrated London News* (18 August 1877).

and finest specimens of apes that have ever been seen out of the forests of their native wilds', he concluded that they were members of the chimpanzee family. Not to be outdone, the primates' manager, J. A. Bailey, now transformed Chiko's species as well, declaring that 'I see no good reason why the two I own should not be gorillas.'[20]

Another chimp 'gorilla', also imported from Lisbon, was exhibited to great acclaim in 1894 at Austin & Stone's museum in Boston as a bigger and stronger rival to Chiko. Chiko himself died of green apple poisoning on 27 July 1894, his body being stuffed for New York's American Museum of Natural History.[21] This left Johanna as supposedly the only gorilla in captivity in America. With no chimp husband to blow her cover, Johanna was now shamelessly promoted as 'the big and intelligent gorilla who is widow of Chiko, the other big gorilla that died'.[22] It was as 'Johanna, the educated gorilla' that she was taken on tour to London in the winter of 1897–8 for a season at the Olympia Theatre, where she amazed audiences with her dress-wearing, knife-and-fork-eating and cigarette-smoking antics.[23] After a surfeit of these and other charades, Johanna died in October 1900.

The first true live gorilla to arrive in the U.S. was delivered to Boston Zoo in May 1897, but survived only a few days. After three gorillas consigned to it from London and Hamburg had also died in transit, and their first import to arrive alive in 1911 survived only a few days after delivery to the zoo, in 1912 the New York Zoological Society commissioned Richard Garner, who had undertaken landmark gorilla observations in the Congo in 1893 (see chapter Two), to trap baby gorillas in Africa and, prior to their transatlantic shipping, 'to remain in that country for two years after specimens were captured in order to accustom them to their white captor as companion, and teach them to eat more civilized food than the acrid food of their native jungles'.[24] In the course of twenty years of field-based research Garner had tended to twelve captive gorillas in Africa, some of whom had survived for several months under his care. For the Bronx Zoo he now procured two specimens, a young female of around three years whom he christened Dinah, and a nine-month-old male named Don. Unusually, Garner seems to have procured these infants by entrapment rather than the more usual method of shooting their parents; the Bronx Zoo's director William Hornaday noted in 1915 that the French government's prohibition of gunpowder use in their Congo territory prevented Garner's importation of 2.3 kg (5 lb) of ammunition intended for local gorilla hunters.[25]

The infant Don died after spending three months at Garner's base camp at Fernan Vaz in the French Congo. His companion,

Dinah, following more than six months of acclimatization and bonding with Garner, was shipped to New York in August 1914. On the crossing to America she occupied Garner's cabin, with Garner himself dossing down in the tool house on the ship's deck originally booked for the primate, and on constant duty as her cook and valet – a situation that aroused considerable press interest.[26] As the sole gorilla in New York, and 'the last wild animal to come out of Africa prior to the complete embargo of war', Dinah retained the attention of the syndicated press after her arrival, becoming America's first true celebrity gorilla.[27] Dinah was to pose from life for two sculptures during her time in New York, the first a bronze portrait bust by Eugenie F. Shonnard, and the second a life-size 37-inch (94-cm) bronze of her seated in contemplative repose by Eli Harvey, an *animalier* sculptor from Ohio who had studied under the celebrated Emmanuel Fremiet in Paris in the 1890s.[28]

When disembarked in New York, this three-year-old gorilla was described by the Bronx Zoo as 'cheerful and affectionate, very lively and full of playfulness, and her appetite is everything that could be desired'.[29] After Dinah's capture, Garner had made what he described as the 'first methodic attempt to change the dietary habits of the gorilla'. Feeling the notion of gorilla vegetarianism to be 'an egregious error', he supplemented her diet with uncooked ham, roast meats and ragout of chicken.[30] This regimen was continued at the Bronx Zoo, where Dinah's vegetarian snacks were outnumbered by her daily intake of raw eggs, milk, and a full lunch delivered hot by a dinner-carrier from the Rocking Stone Restaurant – roast beef, grilled chicken or lamb with lashings of mashed potatoes, gravy and bread. Given these factors, and with Dinah's exposure to her first Manhattan winter, it is amazing that she survived eleven months in the U.S. before expiring.

For the more than two million people who visited the Bronx Zoo in 1914–15 Dinah provided a first and thrilling gorilla experience.[31] Just a few months after her arrival, however, in the depths of winter, zoo visitors were witness to an already ailing Dinah being wheeled in the open for several hours a day in a high-sprung baby carriage, 'dressed like a French doll, with a lace cap, and tucked in with a fur rug'.[32] Black and white footage of her taking in hopefully recuperative fresh air in this winter attire was filmed by zoo curator Raymond Ditmars and screened to fascinated audiences at Wallack's Theatre. An initial diagnosis of infantile paralysis was later changed, after testing of the gorilla's spinal fluid, to one of locomotor ataxia. Despite last-minute forced feeding and the provision of a small bull terrier as a playmate, the diminutive creature died on 1 August 1915; her death was deemed caused by malnutrition.

The fourth gorilla to be brought alive to the USA attained brief celebrity through his connection with a popular circus. Captured in the jungle as an infant (possibly just under three years old) and taken first to Le Havre by a French ship's captain, the gorilla named John Daniel was subsequently sold to an animal dealer and then transferred to England in July 1918. Here he entertained visitors to the pet section of the fashionable Kensington department store Derry & Toms, where he was first seen by Alyse Cunningham in the early winter of that year. When the animal developed influenza symptoms in December 1918 he was purchased from the store by her nephew, Major Rupert Penny. His aunt Alyse's residence at 15 Sloane Street, London, was to be the infant's home until March 1921. Over a six-week period the small gorilla was toilet trained; and, in February 1919, he was given the run of the house. Flourishing on a diet of milk, fruit, jelly, rose petals and tree buds, he soon learned to open doors and windows, turn on lights, pull his chair up to the table at mealtimes and obtain drinking water from a tap (which he always turned off). Cunningham maintained that a sensible feeding regimen, coupled with an intuitive understanding of ventilation and temperature control, created an ideal environment for the animal's rapid and stable development.

During the summers of 1919 and 1920 John Daniel was loaned to the London Zoo three days a week in fine weather, travelling back and forth in a taxi with Alyse Cunningham. Outings were also made to the village of Uley in Gloucestershire, John Daniel travelling by train without any collar or restraint. *The Times* commented upon how the young gorilla's owners had 'treated it as much as possible like a human child, and it is a testimony to their kindness and care that it has remained in good health for two winters. Gorillas are undoubtedly delicate animals in this country, and very few of the importations have lived more than a month or two.'[33]

In January 1921, when John Daniel had grown too large and strong to be safely kept at home any longer, Cunningham and Penny reluctantly decided to part with their charge. Cunningham later maintained that she thought she was selling her gorilla to a wealthy Florida-based animal collector.[34] For his part, the broker of the 1,000-guinea sale, a circus employee named John T. Benson, later confessed to having duped Cunningham, purchasing John Daniel for the Ringling Brothers' Circus, who now brought their formidable publicity machine to bear on their new acquisition. Daniel was claimed to have sailed across the Atlantic in a deluxe stateroom and was greeted on arrival in New York by a brass band, for which grand occasion he was dressed in a sailor suit.[35] Slotted immediately into the Ringling Brothers' annual engagement in Madison Square Garden, John Daniel was advertised as 'positively the only gorilla in captivity'. Isolated, confined to a small space and gawked at by thousands of leering strangers, the still juvenile gorilla became sick and melancholy. As the *New York Times* reported:

> John showed that he missed Miss Cunningham and seemed humiliated when he was assigned to a cage on his arrival here a few weeks ago. He refused food and would crouch on his iron bed away from the front of the cage and cover himself with his blanket to hide from the crowds.[36]

While John Ringling urgently paid for Alyse Cunningham's passage over from London in a bid to revive the gorilla's morale, circus staff moved the ailing John Daniel to a room on the top of Madison Square Garden, where he could see the sky and breathe fresh air. Despite these efforts the poor creature expired from pneumonia on 17 April 1921; a grief-stricken Cunningham proclaimed that he had died of a broken heart after being caged for the first time in his cognisant life. His body was given to the American Museum of Natural History, where it aroused strong interest from scientists and surgeons alike. Some ten neurological, orthopaedic, dental, skin and anatomical specialists requested samples, while a Mr Blashke prepared the animal's skin for taxidermy and subsequent display in the museum.[37]

Alyse Cunningham went on to acquire another, slightly older infant gorilla from West Africa, treating it in a similar manner as a human child rather than an ape. When in April 1924 this second gorilla, John Sultan, was also brought to America for a stint with the Ringling Brothers, this time he was personally accompanied by Cunningham. Carl Akeley was among those on hand to greet the new gorilla, whom the press quickly dubbed 'John Daniel 2nd', when his boat docked in New York. Hailed once again by the circus as 'the only living gorilla in captivity', John Sultan was housed at the Hotel McAlpin during his Manhattan sojourn, and was successfully inspected by a team of scientists at the American Museum of Natural History.[38] Back in London the following year, this second gorilla-child was once again taxied back and forth to the London Zoo for daily guest appearances.[39] He was living proof, said *The Times*, that 'when caught young and treated kindly [gorillas] show intelligence of an order much higher than that of the chimpanzee and the orang, and in its quality closely approaching that of mankind'.[40]

When John Sultan died in July 1926, only one gorilla reportedly remained alive in captivity outside Africa: Miss Congo, the young female brought back to the U.S. by Ben Burbridge. Such was the desirability of having a live gorilla on show that the Al G. Barnes circus toured another gorilla impostor across the U.S. at this time. This was in fact the celebrity orang-utan Joe Martin, star of more than twenty animal comedies in the late 1910s and early 1920s (often billed as a chimpanzee). According to animal trapper Frank Buck, Joe had been forcibly retired from the movie industry after attacking several people upon reaching sexual maturity; he travelled with the Al G. Barnes Show as a caged exhibit, promoted as 'Joe Martin the famous gorilla man'.[41]

In 1915 Bronx Zoo director William Hornaday had brooded that 'there is not the slightest reason to hope that an adult gorilla, either male or female, ever will be seen living in a zoological park or garden'.[42] Up to this point all of the gorillas brought to Europe or America had been infants whose first trauma had usually begun with the shooting of their parents, and who subsequently fell into a physical and seemingly also spiritual decline. 'There seems to be no particular ailment from which they are suffering', lamented animal writer Ellen Velvin in 1914, 'no cold or fever, nothing but intense home- or heart-sickness; and there is no doubt whatever to my mind that they grieve themselves to death.'[43] For the German zoologist Alexander Sokolowsky, writing in 1908, 'before all else spiritual influences undermine the health of captive gorillas'.[44]

In the late 1920s significant changes occurred in the care of captive gorillas, and their longevity increased dramatically. At

Female Western lowland gorilla.

the Bronx Zoo, for example, veterinarian Charles Noback challenged the notion that gorillas should be kept constantly at tropical Congo temperatures, allowing them to move outdoors in all climes; one gorilla in particular, Janet Penserosa, who was brought to the zoo in 1928 in a sickly condition, amazed staff by adapting to the harshness of New York winters, playing in the snow and nibbling on icicles with no ill effects.[45] In addition,

Noback sought to address the need of infant gorillas for affection and companionship by pairing them with 'sibling' chimpanzees of a similar age.

The acquisition of a pair of young gorillas in August 1932 prompted the London Zoo to address the problem of housing these still rare creatures. The Russian architect Berthold Lubetkin was commissioned to design a striking modernist gorilla complex, a circular construction containing in its northern half living quarters that were enclosed, insulated, air-conditioned and lit by clerestory windows, with an open-air cage at the southern end. A metal screen could be quickly revolved to completely encase this outer area, protecting the gorillas from sudden inclement weather. This became the new home of the highly popular Mok and Moina, a pair of gorillas from the French Congo who had already been cared for there by a French official for two years before shipment to London at a relatively mature age (they were seven and nine years old respectively when they left their temporary digs in the lemur house and moved into their grand new residence in April 1933). They acclimatized to London better than their predecessors, although Mok died in January 1938 from unspecified causes.[46]

The London Zoo's next gorilla resident was acquired by means of an exchange with the Paris Zoo. In late 1947 a tiny male gorilla, weighing only just over 10 kilograms despite his age of around two years, was delivered clinging to a hot-water bottle for warmth. Named Guy, after his arrival on Guy Fawkes Day (5 November), he was reported to respond more readily to French than English speech – not surprising, given that this diminutive lowland gorilla had spent six months in Paris after his transportation to Europe from the French Cameroons. Over the next 31 years of his residency at London Zoo, both Guy's size and celebrity were to rapidly increase, his weight climbing to some 200 kg (30 stone) and his fame escalating to even greater levels. Guy's stay in London was not entirely without incident, as he was to bite zoo visitors twice over the years: a small boy in 1956 whose mother lifted him over the safety barrier so he could feed the gorilla; and a 59-year-old man in 1960 who climbed across a 3-foot-high protective barrier so that he could stroke the animal through the bars of his cage. Guy's celebrity survived these setbacks, and weathered too the zoo's importation of a rival young male mountain gorilla in May 1960.[47] When Guy died suddenly in July 1978, of heart failure during an operation being performed to repair teeth rotted by decades of sweets fed to him, the zoo's Annual Report described him as 'the best known and most popular animal in the Zoo – both nationally and internationally. Many thousands of photographs as well as films, portraits and sculptures portrayed his physical magnificence.'[48] After his death, Guy's body was placed on display at London's Natural History Museum; his testes contributed to the research for a paper on 'Testicular Atrophy in Captive Gorillas' in 1980; and artist William Timym was commissioned to create an over-life-sized bronze statue of the gorilla that was installed at London Zoo in November 1982.[49]

Notable imports to the U.S. at this time included Susie the 'zeppelin' gorilla and Bushman, the Lincoln Park Zoo's burly mascot. A lowland gorilla, Susie was captured as a six-month infant in the Belgian Congo, transported to the French Riviera for a twelve-month period of acclimatization and then relocated to Germany. She flew to America as a first-class passenger, checked into Cabin A of the *Graf Zeppelin*, which left Friedrichsafen on 16 May 1929. Her maiden voyage was to be traumatic, as it was equally for the *Graf*'s other nineteen passengers and 40 crew, when four of the dirigible's five engines broke down and the ship was buffeted helplessly for a full day and night by mistral winds. As the airship slanted at an angle of 45 degrees, a journalist on board wired to the world, not surprisingly, that the gorilla was ill. Susie's fame had begun. The ship's commander, Dr Hugo Eckener, managed to finally steer the *Graf* to safety at Cuers, 12 miles north of Toulon in the south of France; and, as the *Graf*'s engines were prepared over the ensuing weeks, the legend of Susie 'the flying gorilla' grew across the Atlantic. 'The adventure of traveling between Europe and America by air is given to few,' mused Michigan's *Ironwood Times*,

> but with a gorilla for a companion it assumes a newer and more romantic aspect than ever before . . . Perhaps 'Susie' will run amok, raging the length of the *Graf Zeppelin* biting the propellers off and casting integral parts of the engine overboard.[50]

The *Graf* resumed its journey on 31 July, this time experiencing a smooth passage on its five-day flight to America. Another journalist, Herbert Siebel, now wired from on board that

> while the world is studiously wondering about us, we are singing and telling stories and having a good time . . . even the gorilla seems to have entered into the spirit of the thing and seems to insist on having company all the time. She is quite affectionate.[51]

When the *Graf* finally arrived at Lakehurst, New Jersey, on 4 August 1929, it was greeted by 100,000 cheering spectators. Later that day the dirigible cruised over Manhattan, accompanied by an escort of smaller planes. After disembarking in the Big Apple, Susie was the first gorilla to be toured widely across the U.S. and Canada, moving between various zoos and circuses.[52] In 1931 Susie was acquired by the Cincinnati Zoo where, under the guidance of trainers William and Caroline Dressman, she learned to sit at a table and use eating utensils and napkins, being subsequently advertised as the world's 'first and only trained gorilla'. Her special bond with the Dressmans doubtless contributed to her remarkable longevity, for Susie was to survive at Cincinnati to the grand age of 22 years, making her, on her death in October 1947, at a weight of some 204 kg (450 lb), both the largest known female and the oldest gorilla in captivity. Markers of Susie's status in American popular culture included her annual and well-promoted birthday party, and her receipt in 1936 of a Shelvador electric refrigerator from the Crosley company. Publicity photographs showed her reaching through the bars of her cage towards bananas and celery in its deliciously cool interior, with its new-fangled inner-door shelves filled with apples. Like other celebrity gorillas, Susie was unceremoniously mounted and stuffed after death; her taxidermied body was stored at the University of Cincinatti, where it was destroyed by fire in 1974.

Bushman, who was purchased as a two-year-old infant from an animal dealer in 1930 by Chicago's Lincoln Park Zoo, grew to be, at 248 kg (547 lb), the most famous zoo gorilla in the U.S. at the time of his death aged 23 in 1951. Apart from his enormous size, he was celebrated for his comparative geniality and, like Susie, was given an annual birthday party. After sports-mad Chicagoans warmed to his training exercises as a juvenile with keeper Eddie Robinson, tackling and passing footballs on Lincoln Park's lawns, Bushman's fame went pre-war America's equivalent of 'viral'. Profiling his fourteenth birthday in 1941, as the most popular of the eleven gorillas then in captivity in U.S. zoos, *Life* magazine logged an annual visitation for Bushman of 3.5 million spectators.[53] Despite the need to install plate glass before his cage to prevent him from throwing things at his audiences, he retained his blue-chip rating for two decades, and was acclaimed by *Time* in 1950 as 'the best known and most popular figure in Chicago'.[54] When he experienced heart problems in the summer of that year, a quarter of a million people filed past his cage in a single week to pay their respects. Following Bushman's death on New Year's Day 1951, he too was stuffed, and then placed on display at Chicago's Field Museum.

Concurrent with Bushman's rise to fame as the U.S.'s premier zoo primate came the ascendency of Gargantua. Unlike Bushman, Gargantua always had to earn his living. The Gargantua story began in 1931, when the captain of an African trade ship brought a tragic little creature from the Cameroons to his friend Gertrude Lintz, the wealthy wife of a Brooklyn doctor. This eighteen-month-old gorilla baby, whom Lintz christened Buddha (later Buddy), had been the victim of a nitric acid attack at the hands of a vicious sailor during his transatlantic voyage. The subsequent scarring left the infant with what Gargantua biographer Gene Plowden has described as 'the pronounced and permanent sneer that was to become his trademark – probably the most valuable animal sneer in recorded history'.[55] Under the care of Mrs Lintz and her animal assistant Richard Kroener, Buddy recovered and then thrived in relative freedom for the next six years, growing to a body weight of hundreds of pounds. Then, after he crept into her bed one night when frightened by a loud thunderstorm, Mrs Lintz realized it was time for Buddy to move on. In December 1937 Buddy, accompanied by his keeper, Kroener, was sold to John Ringling North, who was struggling amid a family power vacuum to gain control of the Ringling Bros. and Barnum & Bailey Circus conglomerate that his late uncles had formerly managed.

Renamed Gargantua after the Renaissance author Rabelais' gigantic character, Buddy was soon to become synonymous with superhuman size and strength for millions of grassroots Americans who had never heard of the French humanist who first coined

Gargantua Close-up, Showing His Human Eyes in a Reflective Moment, postcard, *c.* 1940.

this legendary name. It was the legendary publicist Roland Butler who expanded the gorilla's moniker to Gargantua the Great and, despite his largely placid nature, decided on the basis of his facial scarring to market him as 'the most terrifying creature on earth'. The spin worked. According to grand ringmaster Fred Bradna, Ringling invested $80,000 in a spectacular parade designed by Broadway guru Charles LeMaire of Ziegfeld Follies fame, and featuring renowned animal trapper Frank Buck as well as the newly branded Gargantua – who was wheeled out during the circus's season of 1938 at New York's Madison Square Garden in, as *Time* magazine cooed, 'a heavily barred, thickly glassed, air-conditioned wagon drawn by six white horses'.[56] By the following season Ringling's investment had paid off and Gargantua was a celebrity fixture, the supposed 'only full-grown gorilla ever seen on this continent', rescuing the circus from impending financial disaster.

Roland Butler had a field day in the 1940s, marketing Gargantua as the most violent creature ever brought before the American public. Vividly graphic coloured posters were plastered across the country at a cost of $50,000 per year, depicting a snarling gorilla whirling an African tribesman through the air and pursued by a gang of spear-wielding warriors. 'One of showdom's most famous villains', he had his own line of merchandise for sale at the circus, as well as endorsing products as diverse as Corby's Canadian whiskey and Ryerson Steel ('"Greatest Show on Earth" depends on Ryerson'). The gorilla's fame spread as far abroad as Cairns in northern Australia, where the local *Post* happily aped Butler's spiel that the animal's strength was equal to that of 27 men.[57] Gargantua's cage, a 20-foot-long steel and plate glass climate-controlled box constructed at a cost of $20,000 by the Carrier Corporation, became as celebrated as its occupant, and was paraded around the ring at every performance.

Gargantua Close-up, Showing His Human Eyes in a Reflective Moment

Ringling Bros. and Barnum & Bailey Circus, Sarasota, Fla

Gargantua the Great, poster for Ringling Bros. and Barnum & Bailey Circus, 1938.

It is difficult to know from this distance whether journalists genuinely believed in publicist Butler's tales of Gargantua's evil potential or were playfully complicit in beating up this angle in their coverage of the Ringling Circus's star attraction. *Life* magazine felt that he lived for one purpose – murder – declaring that 'his mind, cunning, treacherous, wanton, is like that of a murderous paranoiac'.[58] Popular author Robert Lewis Taylor rivalled Edgar Allan Poe with his breathless account:

> There are eleven or twelve other gorillas behind bars with circuses or zoos at present, all of them having pretty risky dispositions, but none, according to animal men, has either the dedicated homicidal bent or the explosive power of Gargantua . . . Nobody has been in the gorilla's cage for more than ten years; neither has he been out of it . . . His conduct, outside the cage, would be similar to that of Mr Hyde after the separation from Jekyll. He would walk abroad, largely on his knuckles, bent on destruction.[59]

Protected within his air-conditioned cage, Gargantua toured successfully with the Ringling Bros. and Barnum & Bailey Circus for twelve years, even making the crossing to London in the winter of 1939 for a season with the Bertram Mills Circus at Olympia. Butler kept the animal's publicity machine well-oiled, famously organizing a mock romance for him with an eight-year-old female gorilla, M'Toto, who had been raised in Cuba. Their wedding ceremony, on 22 February 1941, garnered enormous press attention despite Gargantua greeting his new 'bride' with disdain. Thereafter M'Toto travelled with the circus as well, isolated within her own cage, while Ringling clowns dressed as bride and groom gorillas parodied their dysfunctional relationship for delighted crowds. Despite these theatrics, John Ringling North's latest stunt was seen to have a potentially serious side, Australia's Adelaide *Mail* reporting that 'if he is successful, he will have carried out the first successful experiment in mating gorillas in captivity'.[60]

When Gargantua finally died at the age of twenty after spending a dozen years 'in a glass and steel prison, with no unfiltered air, no outside noises, and no chance of escape', during which time he had been seen by more than 40 million visitors to the Ringling circus, news of his passing hit the front pages of newspapers across the U.S.[61] Gargantua's skeleton was donated to the Peabody Museum at Yale University, where it remained on permanent display until the 1970s, and is still seen occasionally in special exhibitions.

The struggle for survival faced by the world's first celebrity gorillas became less arduous in the 1950s, once zoos had finally come to terms with the dietary, climate and social conditions necessary for the species' long-term survival in captivity. Another breakthrough occurred in the second half of that decade when the first gorilla was born in captivity in 1956 at the Columbus Zoo in

Ohio. This landmark event was followed by a sudden surge in infant gorilla births in zoos across the world in the 1960s. These births made a welcome contribution to lessening the incidence of legal and illegal seizure of gorillas from the wild, though they did not however stamp out this trade. By 1983 primate specialist Alan Dixson was able to report that of the 467 gorillas held in zoos and research centres worldwide in 1978, 102 had been born in captivity.[62]

In Lucy M. Boston's classic children's novel of 1961, *A Stranger at Green Knowe*, two displaced orphans bond in a friendship of outcasts. Stolen from the Congo forest after the murder of the elders in his group, the infant gorilla Hanno has been consigned to 'solitary' in London Zoo. Ping, a Chinese refugee, becomes Hanno's secret companion after Hanno, having escaped from Regent's Park, takes refuge himself in the grounds of Green Knowe Manor, where Ping has been billeted for the holidays. What follows is a discussion for teenage readers of the issues of the disappearance of native habitats in a modernizing Africa, impending species extinction, and the binary nature of the zoo as both prison and Ark.[63]

Today's celebrity gorillas continue to be the focus of enormous popular and media attention in the cities where they are held in zoos and animal parks, and their births and deaths are occasions for immense emotion and press coverage. The passing in 2008 of London Zoo's Bobby, the star attraction at that institution's £5.3 million Gorilla Kingdom enclosure, was front-page news. Noting that 'Bobby leaves behind his family of keepers and three female gorillas', a zoo spokesperson asked that 'everyone respect the feelings of staff during this difficult time'.[64] The ability to maintain and breed gorillas in captivity in perpetuity brings other problems, however, such as the inability of institutions to replicate the species' naturally nomadic feeding and nesting behaviour,

Peter Boston, cover illustration for Lucy M. Boston's *A Stranger at Green Knowe* (1961).

as well as the long-term psychological effects upon the animals of confined incarceration and enforced human habituation.

In the late 1980s the British gambling tycoon John Aspinall, a passionate naturalist who founded Britain's Howletts Wild Animal Park in 1957 and Port Lympne Zoo in 1976, began funding the Projet Protection des Gorilles in Gabon and the Republic of Congo. Leasing sanctuaries in these countries, the Aspinall Foundation established them as orphan gorilla camps, caring for infant and juvenile gorillas whose family groups had been wiped out by poaching and civil war, before re-releasing them into the wild.[65] Since 1987 dozens of orphan gorillas have been given a second chance in this manner, their survival rates being monitored as greater than 80 per cent. More extraordinarily, the Aspinall Foundation has begun reintroducing into Africa gorillas that were born and raised at the Aspinall properties in the UK. The reunion in 2010 after five years in the wild of Kwibi, a Western lowland gorilla transported to Gabon in 2005, and Damian Aspinall, who had raised him in the UK after his birth at Howletts Wild Animal Park in 1998, was filmed for Animal Planet's *Gorilla School*. A promo for this sequence of the programme subsequently went 'viral' on the Internet, attracting millions of viewers to witness this new episode in human-gorilla interaction.[66]

4 Sex and Crime

> It is to be hoped that the gorilla will, before long, be classed among the species that have become extinct; for although we very emphatically deny that he is 'a man and a brother', it would be pleasant to have his claims upon us removed as fast as possible.
>
> Norman Macleod, 'The Gorilla', in *Good Words for 1861* (1861)[1]

Throughout the seventeenth and eighteenth centuries in Europe and North America, tales of lustfully marauding primates paralleled serious investigation of chimpanzees, orang-utans and finally – and most luridly – the gorilla. At the Paris Salon of 1887, Emmanuel Fremiet unveiled a second life-size sculpture, *Gorilla Carrying off a Woman*, that commented upon recent and controversial advances in the study of archaeology and prehistory. Presenting a highly dramatic life-and-death scenario from the Stone Age, Fremiet's second gorilla is surrounded by a pack of flesh-eating hunters – visible only to our imagination – from whose midst he has snatched a curvaceous woman. A predator herself who wears a gorilla's jawbone as a hair adornment, she is clutched to his breast in an embrace as erotic as it is violent. The gorilla's back is pierced by a large arrow or javelin, and in self-defence he carries a large chiselled rock that resembles a Palaeolithic cutting tool, suggesting that he has developed a degree of cognitive intelligence. Although critical opinion was hotly divided over the merits and morality of Fremiet's *Carrying off a Woman*, he was now praised rather than shunned by the Salon jury. The sculpture was awarded a Medal of Honour, the first medal the artist had received at the Salon for 36 years. *Gorilla Carrying off a Woman* went on to win another gold medal at the International Fine Arts Exhibition in Munich in 1888, and was shown again at the Expositions Universelle in Paris in 1889 and 1900.[2] Primarily by means of its wide reproduction via small bronze copies, as well as the more affordable media of engraving and photography, Fremiet's sculpture entered the public consciousness in the late nineteenth century as one of the defining images of its time.

Fremiet's sculpture crossed into North American popular culture in particular by means of its importation into New York as an animated wax replica and the subsequent protests of public

Emmanuel Fremiet, *Gorilla Carrying off a Woman*, life-size plaster, shown at the Paris Salon of 1887.

morals campaigner Anthony Comstock, founder of the city's Society for the Suppression of Vice and author of the polemical tract *Morals Versus Art* (1887). In November 1893 Comstock attempted to have the Grand Jury suppress any advertisement of the new exhibit at Manhattan's Eden Musée wax museum, an enormous replica of Fremiet's gorilla sculpture that was 'the latest thing in waxworks', in which the ape's 'eyes were rolling fiercely . . . as he looked down at the victim clasped in his long,

Emmanuel Fremiet, *Gorilla Carrying off a Woman*, 1887, reduced-size bronze.

Handbill for the 1921 United States release of *The Golem*, 1920.

hairy right arm . . . his awful teeth were gnashing by clockwork and he was altogether horrid to the crowd that gaped at him'. What Comstock objected to especially were the 150 'life-size lithographic pictures' depicting Fremiet's *Gorilla Carrying off a Woman*, 'much more lurid than the group itself', that the Eden Musée's manager E. J. Crane had printed as 28-sheet posters and plastered 'all over New York, Brooklyn, Jersey City, Hoboken, Mount Vernon, Yonkers and adjacent towns'.[3] As usual, this attempt at censorship served only to more widely promote the offending phenomenon, such that a few weeks later the *Evening World* newspaper commented upon how 'The wax group entitled "The Gorilla and the Woman" that has been so completely advertised by the interference of Anthony Comstock attracted much attention at the Eden Musee yesterday.'[4] This exposure helps explain why Fremiet's sculpture permeated visual culture so deeply in the U.S., being recycled almost verbatim in the marketing campaign for Carl Boese's masterpiece of German Expressionist cinema *Der Golem* on its U.S. release in 1921 (with Fremiet's nude figures now clothed and the Golem's head grafted on in place of the gorilla's), as well as for the exploitation film *Ingagi* (1930), whose advertisements featuring gorillas and naked women openly acknowledged their debt to 'this famous sculptural group by Fremiet'.

Fremiet had created an enduring image that encapsulated, in colossal three-dimensional form, fears of the primal jungle, of the still misunderstood and dreaded gorilla, of Darwinian theory and its disturbing suggestions that humanity was descended from such base creatures, and above all of unbridled sexuality, particularly of interspecies or inter-'race' sexual contact, imaged through the white woman ravished by the non-white 'other'. This trope surfaced constantly in films from the 1930s onwards which heavily mined these fears of bestial passion and 'miscegenation' through depictions of apes – and specifically gorillas – as either violently homicidal or curiously licentious (the Fay Wray-sniffing Kong in RKO's legendary *King Kong* of 1933), with the gorilla embodying chaos repressed by civilization in both cases.

Part of the blame for the evil typecasting of the ape as killer can be attributed to the 'butchery without motive' of the orang-utan in Edgar Allan Poe's popular *Murders in Rue Morgue* (1841), which was generally imaged as a gorilla in subsequent film adaptations. Short stories, novels, plays and films portrayed the gorilla as murderous and aggressive. In 1944 Disney's animation *Donald*

Duck and the Gorilla, in which Ajax the killer ape has escaped from a zoo to terrorize Donald and his nephews before being subdued with tear gas, sent up this vast genre of stereotyped gorilla-maniacs.

In April 1925, after a lacklustre tryout upstate, Ralph Spence's play *The Gorilla* suddenly became a smash hit on Broadway and quickly moved across the Atlantic to the London stage as well. A

"ONE OF THE FUNNIEST AND MOST THRILLING SHOWS EVER SEEN ON BROADWAY"

DONALD GALLAHER and JAMES W. ELLIOTT present

The GORILLA

A THRILLING, CHILLING, KILLING, MYSTERY BY RALPH SPENCE

Staged by Walter F. Scott

Newspaper advertisement for Ralph Spence's play *The Gorilla,* 1925.

spoof of the mystery genre, Spence's play was set in a city terrorized by a series of murders committed by a lunatic of superhuman strength and ape-like appearance. Featuring bumbling detectives, an escaped gorilla and a maniac, it delivered comedy and suspense in equal measure. A key to its success lay in Spence's denouement, when 'the gorilla leaped off the stage and growled its way up and down the aisle, frightening many of the women, and causing some of them to faint.'[5] The stage success of *The Gorilla* led to its being brought quickly to the screen, firstly as a silent feature in 1927 and again with sound in 1930. The thrills were kept alive in the first film by Charlie Gemora's personification of the gorilla, whom viewers found to be 'a weird, menacing looking ape, and every time he appears the crowds shriek as if he were right in the audience'.[6] Spence himself penned the screenplay for the talkie version; and both films maintained suspense by altering their endings from that seen by visitors to the play's year-long run on Broadway. For months, as the films criss-crossed America, hoardings and newspapers were plastered with images of gorillas carrying off a scantily clad woman or strangling Spence's hapless but loveable sleuths.

What gave these film versions of *The Gorilla* a strange topicality and a serious grounding in the popular imagination despite their superficial frivolity was their release at a time of bizarre and relentless sex crimes in the U.S., committed by a series of lunatics whom the press uniformly called Gorilla Men.

On 9 June 1927 the front page of the *Los Angeles Evening Herald* bore the headline 'Gorilla Man Brands L.A. Beauty', signalling one of that year's strangest incidents. The case involved a young actress, Doris Dore, 21, who answered knocks on the door of her apartment at 2 a.m. only to find herself seized by the 'terribly long gorilla-like arms' of an unknown assailant who proceeded to brutally carve the letter K onto her forehead, chin, breasts and arms with a razor blade.[7] Grotesque as this crime was, it paled in comparison with the exploits of the necrophiliac 'Gorilla Man' murderer who terrorized the U.S. and Canada for eighteen months between 1926 and 1927.

In January 1927 newspapers had excitedly reported the arrest of a suspect believed to be the notorious Gorilla Man, who had in the preceding months brutally strangled and then sexually assaulted the corpses of at least eight women in California, three women each in Oregon and Missouri, and two women and a baby in Kansas City.[8] This was a false hope, however. The wrong suspect was in custody and America's first Gorilla Man serial killer went on to strangle at least six more women in the Midwest and in Winnipeg, Canada, before being apprehended. The few reliable eyewitness accounts of the murderer described him as ape-like in appearance, long-armed, shuffling in gait and prone to manic laughter as he fled each new slaughter. 'It was those long arms and sinewy hands, his shambling walk and his animal brutality that earned him his title when his name was unknown', the press noted.[9] When Earle Nelson, an ex-inmate of the Napa State Mental Hospital, was finally arrested as the Gorilla Man on 16 June 1927, his known women victims numbered more than twenty – all of them strangled and raped post-mortem, and some also horribly mutilated and dismembered.

North America's paranoia over these Gorilla Man attacks is perhaps reflected in the trauma experienced by Mrs Hazel Pasley in Los Angeles in 1929 after being hugged on a downtown street by a man dressed in a gorilla suit, who was promoting a new film for the Principal Pictures Corporation. Mrs Pasley argued persuasively in court that the incident 'caused her to suffer a nervous breakdown and that a picture of the man "thrusting his projecting fangs and hideous face" before her had remained in her mind', leading her to suffer 'from delusions that hair like that of the gorilla was growing on her back'.[10]

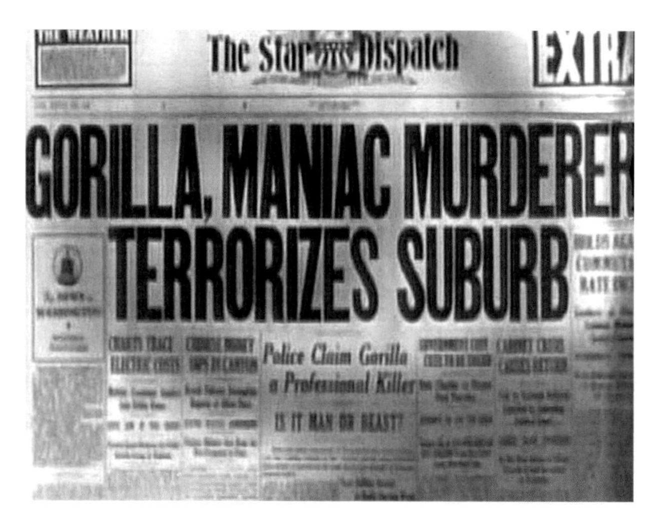

The Star Dispatch

GORILLA, MANIAC MURDERER TERRORIZES SUBURB

Police Claim Gorilla a Professional Killer

IS IT MAN OR BEAST?

Newspaper headline from *The Gorilla* (1939).

When Earle Nelson was executed in Winnipeg on 13 January 1928, while the silent film of *The Gorilla* was still playing to packed houses, this should have been an end to this real-life Gorilla Man carnage. Nelson's atrocities, however, seem to have inspired a series of copycat Gorilla Man sex crimes that plagued the U.S. throughout the 1930s. To cite just a handful of cases: in 1932 a muscular itinerant blacksmith named Gilbert F. Collie was convicted of two Gorilla Man murders in California. He made national headlines again in 1956 when a judge refused his request to be executed and thus end his seemingly interminable life imprisonment.[11] In October 1933 Florence Johnston's naked body was found strangled and eviscerated with gorilla-like force in her New York apartment. This sparked new Gorilla Man killings in San Francisco that ended with the arrest of Florence Johnston's husband for all these crimes in July 1940.[12] Meanwhile, in January 1934 California froze in terror as Collie, 1932's 'Gorilla Man', escaped from the Mendocino State Hospital for the Criminally

Title screen from *The Gorilla* (1939).

Al Ritz with Poe the Gorilla (played by Art Miles) in *The Gorilla* (1939).

Bela Lugosi with the gorilla (played by Emil Van Horn) in *The Ape Man* (1943).

Insane. In San Diego in September 1936 Donald Hazell, that year's '225 pound "gorilla man"', confessed to killing women whenever 'psychic spells come over me and compel me to take a life'. In May 1938 Marion Elliott of Los Angeles survived a blow to the head delivered by a man matching the description of the laughing Gorilla Man who had bludgeoned another woman to death two days earlier. Less than a week later, a woman was bashed to death with a brick in Chicago by a suspect that the newspapers called 'a Negro "gorilla man" who fled through a broken window'.[13]

Hollywood was therefore very much responding to current events when it chose to revisit Ralph Spence's play *The Gorilla* yet again with Darryl F. Zanuck's film of the same name in 1939, starring the popular Ritz Brothers comedy team. Shivers of recognition doubtless passed through audiences when the film opened with a series of sensational newspaper headlines – 'Gorilla Maniac Murderer Terrorizes Suburb' and 'Gorilla Claims Fifth Victim' – shown in montage form over the leering features of a gorilla, a screaming woman and images of police mobilizing for a manhunt. A complex farce was then set in motion within the film, involving a real gorilla from a county fair (named, aptly, Poe) which had been kidnapped and then set loose as cover for the murderous actions of a gorilla-suited maniac. The shadow of appalling recent events underlies this film's apparent levity, as it does Universal's equally farcical *The Strange Case of Dr Rx*, released in 1942, which also revolves around five strangulation murders, and features Nbongo (Ray Corrigan), a wonderfully lurid and frisky gorilla chained up in the mad Dr Rx's basement for no apparent reason other than the false threat of a brain-transplant for the film's detective hero. The real-life Gorilla Man trope also informs the newspapers that shriek 'Ape Man Runs Wild', in Monogram's Bela Lugosi vehicle *The Ape Man* (1943). Here Lugosi, who has been regressing into a hybrid man-ape since experimenting on himself with simian DNA, co-opts his laboratory gorilla (Emil Van Horn) into committing four strangulation murders so that Bela can extract the fresh spinal fluid he needs for his reversion to human form.

A United States haunted by countless gruesome real-life slayings is echoed too in the banner headlines, 'Gorilla Man Still at Large. Slayer Evades Police', that blare out in Warner Brothers's spy film *The Gorilla Man* (1942). In this wartime thriller, John Loder uncovers a secret Nazi stronghold on the British coast. In order to discredit his evidence, Nazi goons commit a series of grisly strangulation murders, framing Loder as their Gorilla Man perpetrator. While *The Gorilla Man* contains no actual gorilla action, its promotional advertising used predatory gorilla imagery to great effect.

While gorillas in zoos became the objects of popular affection, a fear of larger apes remained embedded in public consciousness in Europe and North America. The fear of the killer gorilla was inexorably linked to the fear of the captive gorilla escaped from its confines – circus, zoo or laboratory.

The first escapee gorillas reported in the U.S. were clearly chimpanzee imposters, since their sightings predated the successful transport of live gorillas to the country. Both the gorilla believed to inhabit the woods in South Norfolk in 1894 and the wild man who jumped a fence in front of a stagecoach near Winsted, Connecticut, in August 1895 must have been chimpanzee circus runaways, if indeed they were not just figments of the imagination.[14] The 'gorilla' who 'scattered pedestrians right and left' in Memphis, Tennessee, in 1888 was a chimp from Forepaugh's Circus; and it was another chimp 'gorilla' from Barnum & Bailey Circus (the impostor Chiko again) who invaded the ladies' dressing rooms in Binghamton, New York, in May 1893.[15]

In 1930 a group of farmers in southwest Iowa were convinced that they had all sighted a gorilla absconder. The creature, whose chances of survival in a Midwest winter were surely slim, was supposed to have escaped a year earlier from a circus at Glenwood.[16] Like the menacing Ajax in *Donald Duck and the Gorilla* (1944), escaped gorillas populated cartoons for decades. Disney provided another of the classic characters in this genre with Beppo, who kidnaps Minnie Mouse in *The Gorilla Mystery* (1930). In the 1940s, a 'man eatin' gorilla' who had escaped from a circus was tamed by Li'l Abner; and another escapee ape, Bozo, had run-ins with rotund cops in cartoon advertisements for Cream of Wheat and Pepsi-Cola, among other products. Such was the popularity of the trope that in 1934 Harry Beaumont's railway mystery *Murder in the Private Car* concluded with an escaped circus gorilla that had nothing otherwise to do with the film's plot.

In 1922 Eugene O'Neill's classic play *The Hairy Ape* conjoined escape and murder with the gorilla to dramatic effect. Described by the *New York Times* as 'a bitter, brutal, wildly fantastic play of nightmare hue and nightmare distortion', *The Hairy Ape* shocked audiences in the 1920s with its tale of the hirsute, grubby ocean liner stoker Yank, who becomes alienated after being rejected by a socialite woman as a 'filthy beast'.[17] In the play's final scene, set outside the gorilla cage at the Bronx Zoo, Yank laments to the ape: 'Ain't we both members of de same club – de Hairy Apes? So yuh're what she seen when she looked at me, de white-faced tart! I was you to her, get me? On'y outa de cage – broke out – free to moider her, see? Sure! Dat's what she tought.' When Yank, in despair, breaks open the gorilla's cage, seeking solidarity, the animal brings not comradeship but death, crushing the life out of the sailor before fleeing into the night. O'Neill's play, a controversial success, was transferred to the London stage and also translated into French as *Le Singe velu*, playing at the Théâtre des

Arts in Paris in 1929. Its vivid gorilla murder is echoed in the film *Captive Wild Woman* (1943), when Cheela the female gorilla strangles a man though the bars of her cage, giving rise to more strident cinema headlines ('Circus Handler Murdered by Ape!').

The gorilla as escapee informs Monogram's *The Ape* (1940), a rather badly produced Boris Karloff vehicle whose narrative is set in motion by the strangulation murder of an abusive keeper at Posts Combined Circus and subsequent rampages by the gorilla giant Nabu. Nabu's path collides, against the backdrop of a massive gorilla hunt in rural Red Creek, with that of Karloff's Dr Adrian, whose obsession with curing a young girl's paralysis has tipped him over the edge. Adrian seems to have discovered a paralysis cure by serendipitously extracting spinal fluid from one of Nabu's victims – but he needs more human bodies to complete his experiments. When Nabu invades Adrian's surgery he is killed by the doctor, who flays the gorilla and wears its skin as a Nabu-imposter, terrorizing Red Creek himself while hunting

Ray Corrigan as Nabu the Ape in *The Ape* (1940).

down future victims for his vital spinal juice before being shot dead, dressed in his full ape regalia, by the local militia.

Cinema's most spectacular escapee gorilla action was provided by Twentieth Century Fox's 3D extravaganza of 1954, *Gorilla at Large*, which offered audiences a Technicolor widescreen fantasy of what Ringling Brothers' celebrity gorilla Gargantua's life might have been like had he ever been released from his air-conditioned mobile cage. More frothy-mouthed and down-at-heel chic than his real-life inspiration, this film's enormous gorilla, Goliath, is held captive in a combination circus and amusement park (he is played in a deliciously shabby flokati body suit by George Barrows, a stunt actor who went on to provide cheesy gorilla impersonations for numerous 1960s television programmes, from *The Addams Family* to *The Beverley Hillbillies* and *The Man from U.N.C.L.E.*).[18] Goliath is subjected to publicly ticketed sexual teasings from Anne Bancroft, playing a scantily clad red-sequined trapeze artist who performs barely inches above his lustful, predatory grasp. At the end of a convoluted plot involving diverse suspects and accusations of murder past and present, Goliath escapes his cage with Bancroft in his clutches, scattering theme park patrons willy-nilly and scaling the roller coaster before being shot by state troopers.

While at first glance *Gorilla at Large*'s plot seems utterly preposterous, the 1950s actually witnessed a number of dramatic gorilla escapes. In October 1950, after twenty years of solitary confinement in his glass and steel cage at Chicago's Lincoln Park Zoo, the 250-kg (550-lb) Bushman took advantage of a momentarily unlocked door to his quarters and made his escape on a busy Sunday when the zoo was filled with visitors. When challenged, he bit his keeper Eddie Robinson on the arm and proceeded to create havoc in the monkey house kitchen, where food was prepared for the zoo's primates. Meanwhile, as local papers reported the incident, 'more than 50 heavily-armed police surrounded the building to keep away crowds and block off zoo roads against the awful possibility that the 6-ft-2 giant, who could kill a man with a swipe of his hairy arm, might break out.'[19] Another dozen police armed with rifles and shotguns stood guard within the monkey house for a tense two hours before Bushman was cowed by his innate fear of reptiles and forced back into his cage by the threat of a 2-foot garter snake.

The following year state, county and city police were called in to help search for a 135-kg (300-pound) gorilla that had escaped at the winter quarters of the Hennies Brothers Carnival in Hot Springs, Arkansas. It had the run of the carnival's 38-acre compound for an entire afternoon before being lured back to its cage with foodstuffs.[20] In France George Brassens' popular song 'Le Gorille', released both as a single and as part of his *Mauvaise Réputation* album in 1952, approached the trope with Continental flair. The provocative lyrics of Brassens' song tell the story of a sexually frustrated virgin male gorilla who, having been aroused by women ogling his genitals, escapes from his confines and sodomizes a magistrate. Perhaps not surprisingly, it was banned from French radio upon its first release.[21]

Another gorilla escaped from a circus in Almansa, Spain, in May 1956, frightening passers-by as it roamed the streets of the small town for two hours before being captured in a convent school. Five months later a male gorilla named King Kong made a break for it at the Leo Carroll animal show in Tampa, Florida. This 105-kg (230-lb) creature attacked and mauled his keeper before being driven off by a gunshot fired by a fellow employee. The animal was killed by Deputy Sheriff Lloyd Simmons with a bullet to the head.[22]

In July 1958 an extraordinary gorilla incident occurred in Antwerp when a male ape, Kobus, escaped from his cage while being fed, scaled the zoo walls and terrorized early morning shoppers before climbing up the front of a house. Then, in a manner reminiscent of the flight of Erik the gorilla in Robert Florey's film adaptation of *Murders in the Rue Morgue* (1932), he made his way across a row of rooftops while police evacuated houses and cordoned off the area. Corralled by the local fire brigade's water hoses, he became enraged, smashing chimney pots and roof tiles. Faced with this highly dangerous situation, zoo officials reluctantly ordered him to be shot by police stationed on neighbouring roofs. Toppled by six bullets to the chest, Kobus smashed through two verandahs as he plummeted to earth.[23]

Since the late 1980s there has been a remarkable increase in the number of gorilla escapes from zoos, more than a dozen of these occurring in the U.S. alone. While most of these animals were subdued before reaching public areas, there have been some notable exceptions – such as when a twelve-year-old male gorilla jumped across a 12-foot moat in Los Angeles in 2000 and headed for startled schoolchildren; or when a ten-year-old female scaled a 14-foot wall in Pittsburgh in 2001, to the terror of several hundred zoo visitors. In the most unfortunate incidents a 300-lb gorilla from Boston's Franklin Park attacked a two-year-old girl and an eighteen-year-old woman in 2003 before fleeing into the streets outside the zoo where it was tranquillized by police; Hercules, a 33-year-old escapee male at Dallas Zoo, bit a young keeper 33 times on the face and body in 1998; and another gorilla at the Dallas Zoo, a 300-lb thirteen-year-old male named Jabari, ran amok in 2004, causing the evacuation of hundreds of visitors and seriously injuring two women and a three-year-old boy before being shot dead by Dallas tactical officers.[24]

In a strange echo of Fremiet's sculpture or *King Kong*, a gorilla ran off with a woman at Rotterdam's Diergaarde Blijdorpe zoo in May 2007, dragging her for tens of metres, breaking her limbs and repeatedly biting her. After abandoning her, the gorilla, an eleven-year-old, 180-kg male called Bokito, charged into the zoo's restaurant, sending panic-stricken patrons fleeing in all directions, before being brought down with sedative darts.[25] Bizarrely, Bokito's female victim was known to him; a regular visitor, she claimed to have a special bond with the gorilla, smiling at and kissing him in the weeks leading up to her abduction, across the protective glass wall of his enclosure, unaware that the human smile with teeth bared is read by gorillas as a sign of aggression. Captured on amateur video and still photography, Bokito's abduction of this woman was transmitted around the globe, and 'Bokitoproof' – able to survive a gorilla attack – became the Netherlands' Word of the Year for 2007.

From the first horror films, through *King Kong* and early Walt Disney, to the B-grade horror of the 1950s, the image of the beautiful woman collapsed as helpless victim in the arms of the monster was both scripted as the film's moment of greatest tension and used as a major promotional ploy. The key image of beauty in the grip of the beast was invariably used on the films' lobby cards and posters, and gorillas were one of the most popular culprits in this genre. Cartoon heroines Minnie Mouse and Betty Boop also found themselves carried off by animated gorillas. In Dave Fleischer's *Parade of the Wooden Soldiers* (1933), rockets fired off in celebration of Betty Boop's arrival in a toy store energize a gigantic gorilla rag-doll which goes on the rampage, crushing beneath its feet and dismembering other toys. It runs off with and threatens to decapitate Betty before being subdued and restrained by an army of toy soldiers.

Sexual fantasies involving gorillas predate *King Kong*, notably in cinematic potboilers such as *Ingagi* (1930). Promotional materials suggested that this film showed some sort of missing link (Has

Poster for *Ingagi* (1930).

the Man Ape Been Found?) and naked women (Wild Women! Apparently Half Ape! Half Human!) living with gorillas. It also claimed to feature the sacrifice of a woman to a giant gorilla and posters showed a gorilla carrying a bare-breasted black woman under one arm. The Will Hays Office of the Motion Pictures Producers and Distributors of America banned the film in June 1930. Surprisingly, this was not on the grounds of explicit content but rather because of false advertising, as the gorilla sequences were judged to be made up of stock footage of orang-utans and a gorilla-suited actor. Natural scientists had been particularly vocal in demanding the film be banned from exhibition, believing that it might do moral harm to the gorilla's reputation at a time when efforts were being made to ensure the survival of the already threatened species.[26]

The gorilla as cabaret entertainment was a feature of the Paris nightclub scene as well as cinema elaboration before Kong's appearance on a New York stage. A resident of Winnipeg, narrating his trip to Paris in 1929 for readers back home, described a racy evening spent at the Folies Bergère, where 'A magnificent female athlete . . . strode onto a platform erected in the lap of a gigantic black hairy gorilla, and did muscle and bending movements inspired by the most subtle and sinister poetry of motion.'[27] Gustave Soury designed posters for these 'La Belle et la Bête' evenings at the Folies, showing a nude woman clasped in the grasp of a gigantic leering gorilla.[28] A Spanish film of 1926–7 by Benito Perojo, *El Negro que tenía el alma blanca* (*The Negro with the White Soul*), similarly stages a primitive dance inside a giant gorilla head in a dream sequence portraying fears of black men's sexuality; the woman dreams that she is abducted to the gorilla's mouth by a black dancer.

In Joseph von Sternberg's *Blonde Venus* (1932), Marlene Dietrich lumbers onto the cabaret stage in a gorilla suit amid jungle drums and a chorus line of women with dark makeup and Afros, holding spears and shields. Dietrich strips off the gorilla costume and slips on a blonde Afro wig to perform her 'Hot Voodoo' number. The following year Fay Wray would appear in a blonde wig in *King Kong*, similarly underlining her fragile white beauty amid 'savages' and gorillas. Links between music and sex scandalized and titillated white audiences in the Jazz Age of the 1920s in the U.S. Much of the moral hysteria about jazz related specifically to ideas about black sexuality, in particular the supposedly sexual rhythms of African dance and music. Dietrich's 'Hot Voodoo' number in *Blonde Venus* links dance and music to the gorilla as an expression of primal sexual drives, offering an escape from 'civilization' in the company of a cabaret troupe of 'African' tribeswomen.

In 1931 New York iron and steel heir William Douglas Burden and adventurer and filmmaker Merian C. Cooper developed a project to film mountain gorillas in the eastern part of the Belgian Congo, hoping to build a large enclosure within the gorillas' natural habitat in which to film. As a small child Cooper had been given a copy of Paul du Chaillu's *Explorations and Adventures in Equatorial Africa*, which had seeded his obsession with gorillas,[29] while Burden held a strong belief in the importance of natural history educational films, and five years earlier had funded and participated in an American Museum of Natural History expedition to the Dutch East Indies to capture and record on film Komodo dragons. To capture the gorillas for the enclosure, Harvard zoologist Harold Coolidge recommended the use of gas, but warned that some would no doubt be killed and that a permit would be required. After the *Ingagi* authenticity scandal, as well as questions that had been raised about the ethics of Martin Johnson obtaining a gorilla-hunting permit for *Congorilla*, securing a permit for a film with commercial ends was likely to be

'King Kong' street graffiti, Gifu, Japan, 1994.

difficult. In late 1931 Cooper was invited by David O. Selznick to become an associate producer and executive assistant at RKO. Selznick also promised that RKO would develop the gorilla picture. Cooper saw the stop-motion animation effects that Willis O'Brien was perfecting for a follow-up to *The Lost World*, and realized that location shooting would not be necessary. A giant gorilla living on a remote island, such as that visited by Burden to film the Komodo dragons, could be shown battling dinosaurs: *King Kong* was born.

The film combined an apparent concern for geographic specificity – Skull Island is located 'west of Sumatra' – with the fantastical elements of the giant gorilla cohabiting the island with prehistoric animals and islanders (mainly played by African American actors) whose appearance and material culture are a pastiche of Pacific- and African-inspired elements. The film's Hollywood premiere on 24 March 1933 at Grauman's Chinese Theatre in Los Angeles featured a performance prologue entitled 'A scene in the jungle' by a 'Chorus of Dusky Maidens and African Choral Ensembles', with acts including a 'Dance to the Sacred Ape'.[30]

King Kong has variously been explained as an imperialist parable about the dangers for the primitive of contact with civilization; a cautionary tale about rampant male sexuality and the dangers of interracial intercourse; and an allegory of the return of the repressed in the form of primitive sexual drives. Both the giant gorilla Kong and heroine Ann Darrow (Fay Wray) are objectified and serve promoter Carl Denham's designs to capture the ape for commercial exploitation using the woman as bait. Ann does not just represent beauty, film historian James Snead notes, but 'endangered beauty'; he also points to the sadism in Denham's desire to see her 'molested or at least scared out of her wits by a black ape representing remorseless phallic potency'.[31] When Kong is captured on Skull Island, Denham gleefully declares that 'the whole world will pay to see this'. Kong is a 'noble Savage' who is then exhibited in irons on a New York stage as a spectacle of primitive power. Fatimah Tobing Rony suggests that

> the lineage of *King Kong* should be obvious: the filming, capture, exhibition, photographing, and finally murder of Kong takes its cue from the historic exploitation of native peoples as freakish 'ethnographic' specimens by science, cinema, and popular culture.[32]

Initially Cooper had conceived of Kong's downfall as the result of contact with 'civilization'. He informed his friend William Burden:

Ben Frost, *Love/Hate Pre-War*, 2002, acrylic on board.

> When you told me that the two Komodo Dragons you brought back to the Bronx Zoo, where they drew great crowds, were eventually killed by civilization, I immediately thought of doing the same thing with my Giant Gorilla. I had already established him in my mind on a prehistoric island with prehistoric monsters, and I now thought of having him destroyed by the most sophisticated thing I could think of in civilization, and in the most fantastic way. My very original concept was to place him on top of the Empire State Building and have him killed by airplanes.[33]

Completed two years before the release of *King Kong*, the Empire State Building represented technological progress and modern city life; the fighter planes which would bring Kong down were equally representative of technology and modernity. The juxtaposition of King Kong, the Empire State Building and the planes provides an unforgettable image of the irruption of the primitive in the modern and an allegory of the confrontation between the wild and technological power.

The effect of *King Kong* on popular culture was immediate and profound. In 1933 Disney quickly released the Mickey Mouse cartoon *The Pet Store*, which directly parodied *King Kong*'s finale; as did 'King Klunk', an episode from Walter Lantz's *Pooch the Pup* cartoon series released in the same year. In Disney's cartoon Mickey takes a job at a pet store, where Beppo 'The Movie Monk' is inspired by a movie magazine to impersonate King Kong. Beppo breaks out of his cage, snatches Minnie and climbs to the top of a stack of boxes reminiscent of the Empire State Building. The pet store's population of birds and small monkeys launch an attack against the gorilla, dive-bombing and firing makeshift projectiles at Beppo.

Claudio Onorato, *Mister Empire (sotto il mare)*, 2007, cardboard cut-out.

DELUGE

Gorilla impersonator Charles Gemora with Lou Costello in *Africa Screams* (1949).

The Kong phenomenon continues to resonate: the great ape and his demise have been the inspiration for endless parodies and references in plays, cartoons, television comedy, advertisements, comics, art and cinema, including feature film remakes in 1976 and 2005. The original *King Kong* was re-released in 1938, and again in 1942. It was not surprising then that in Nassour Studio's *Africa Screams* (1949), animal-phobic *faux*-hunter Stanley Livington (Lou Costello) should come across his nemesis in the form of the gigantic Kong-like 'Orangutan Gargantua', named in homage to the Ringling Brothers' star gorilla. It was in the 1950s that the full pop-cultural impact of *King Kong* began to be realized, as the film was screened repeatedly in cinemas, on television and in drive-ins. *Time* magazine dubbed it 'Movie of the Year' in 1952. In this period *King Kong* moved beyond its original horror-adventure film genre when the iconic imagery of a giant gorilla wreaking havoc in a large city spawned similar imagery in several science-fiction classics, including *It Came from Outer Space* and *The War of the Worlds* (both 1953). *Godzilla* (1954) was also heavily influenced by *King Kong*, which enjoyed enduring popularity in Japan, and the big ape went on to feature in Japanese productions *King Kong vs. Godzilla* (1962) and *King Kong Escapes* (1968).

The two closely related remakes of *King Kong* roughly follow the original film's characters and scenario, but with quite different perspectives reflecting their own eras of production. John Guillermin's *King Kong* (1976) focuses more emphatically than the original upon a world destroyed by greed in which the perceived dividing lines between animals and humans have begun to crumble. In Peter Jackson's extravagant and nostalgic *King Kong* (2005) we are invited to witness, as film theorist Barbara Creed has put it, how 'together human and animal celebrate the beauty of the natural world – possibly for the last time'. Jackson's empathetic reworking of Beauty and the Beast is, Creed notes, 'a twenty-first century fairy tale' that 'is not about the traditional quest to find a prince for Beauty; rather it focuses on the tragedy of the animal and, by extension, the death of nature'.[34] King Kong's fascination promises to endure well into the twenty-first century with the Creature Technology Company in Melbourne currently preparing a 7-metre-high motorized Kong puppet that will feature in a new Global Creatures theatre production, *King Kong Live on Stage*.[35]

The image of the powerful and primitive gorilla who is master of his domain gained enormous currency in the U.S. in the first decades of the twentieth century. This was the time when the

Preliminary trials of the animatronic gorilla developed by the Creature Technology Company, Melbourne, for *King Kong*, the musical.

back-to-nature movement first flourished, reflected in the rise of nature preservation societies and the literary taste for accounts of wild animals and adventures, notably Jack London's best-selling *The Call of the Wild* and Ernest Thompson Seton's *Wild Animals I Have Known*. A fascination with brute courage and animal instinct emerged in parallel with the waning of opportunities for their expression in industrial society; contact with nature was considered ennobling and an antidote to the emasculation effected by modern life. The fear of the erosion of virility in industrial society and nostalgia for the world of pioneers made strong through contact with nature brought with it a search for forms of male activity such as hunting which were associated with the wild. As hunting fell from fashion other sports were promoted, often through advertising featuring wild animals and particularly the virile gorilla, whose image combined sex and strength in irresistible combination, associations that continue to be mobilized in advertising today.

The gorilla's enormous power has been equated with a bewildering variety of products. These range from Akron's First World War-era Gorilla Super Tires ('extra heavy and extra tough for heavy duty service'); to today's Cincinnati-based Gorilla Glue Company's range of adhesives; and 'Gorilla, The Strong Yellow Beer', a 7.6 per cent lager brewed by Åbro in Sweden. The gorilla's appealing strength has worked well in cautionary advertising too, as in the 1940s Travelers accident insurance campaign featuring a powerful ape who manhandles a Stradivarius violin. In the late 1970s and '80s 'ape shit' or 'gorilla biscuits' was popular American slang for a strong substance of a darker nature – the powerful hypnotic sedative methaqualone (Quaalude). Its popular use as a recreational drug led to the adoption of this term by the hardcore New York band Gorilla Biscuits. Their self-titled debut album, which became a hardcore hit in 1988, featured a blissed-out simian thug who subsequently became the band's mascot, gracing a variety of band merchandise as well as the tattoos of adoring fans.

The Mighty Gorilla became a promotional symbol for the 58th U.S. Pursuit Squadron during the Second World War in recognition of the courage of its pilots and the guerrilla-warfare tactics they utilized when fighting in the Mediterranean in 1942–4. Echoing this analogy of the gorilla's power with human strength in adversity, the group of ex-prisoner commandoes whom American television gathered together to fight the Nazis for two seasons in 1967–8 were christened 'Garrison's Gorillas'. Strength and reliability epitomized American Tourister's durable luggage in the 1970s, when the U.S. was bombarded with television commercials featuring delightfully wicked taxi driver and bellboy gorillas mauling customers' luggage. 'Beautiful under pressure' and 'built to take it', the American Tourister's products emerged unscathed.[36] Recently an exceptionally tough glass has been developed by Corning. Its name: Gorilla Glass, which the company markets as 'tough, yet beautiful' with advertisements featuring sleekly computer-imaged gorillas.

The softer side of the gorilla was the focus of the highly successful 'Gorilla' marketing campaign used by Cadbury Schweppes to promote Cadbury Dairy Milk chocolate in 2007. This featured a drum-playing gorilla (actor Garon Michael, who had previously aped gorillas in the films *Congo* of 1995 and *Instinct* of 1999) performing Phil Collins's 'In the Air Tonight'. The advertisement's theme was topical, given the recent press speculation about Leah,

Gorilla automobile tyres advertisement, Akron, Ohio, 1918.

Gorilla advertisement for The Travelers insurance, USA, 1949.

the Western lowland gorilla who was observed using tools in the wild in 2005 (see chapter Five).

The term 'gorilla' also readily entered literature and popular culture to denote either hirsute or powerfully muscular men, especially athletes. One of the earliest uses of the term in this context occurs in Robert Ballantyne's boys' adventure novel of 1861, *The Gorilla Hunters*, in which a principal character, the hairy-bodied

Jack Martin, is consistently referred to as such. The handsome African American boxer William Jones, who secured the World Middleweight boxing title twice, in 1932 and 1933, and won 101 out of his 138 professional fights, became known internationally as 'Gorilla' Jones because of his exceptionally long reach in the ring. Glamorous, wealthy and always expensively dressed, Jones met actress Mae West at a New York nightclub in 1928. Their friendship led to Jones becoming West's chauffeur and body-guard after his retirement from professional boxing in 1940. The saucy lines 'Hey, Gorilla. Come here', penned by West for her film *Night After Night* (1932), were a coded message referring to Jones's landing of the middleweight title in that year.[37]

A hirsute professional American footballer and heavyweight wrestler, Arthur ('Tarzan') White, often found himself likened to a gorilla by sports writers. In 1937 journalist Henry McLemore, commenting on White's inaugural game as offensive lineman with the New York Giants, wrote cheekily: 'It is reported that Tarzan turned down a good job in Hollywood (as a stand in for a gorilla) to accept the job with the Giants. He stands 5 feet 8 inches, weighs 220, and is covered with a luxurious coat of fur.'[38] In more recent slang, the term 'gorilla' can be applied locally to the locks of hair on the back of a hirsute man's neck ('Barber, just trim my gorillas').

The cauliflower-eared wrestler and actor Luigi 'Bull' Montana, who had coincidentally appeared as the Ape Man missing-link character in the feature *The Lost World* (1925), liked to amuse the press by confessing that he looked 'more like a gorilla than most gorillas do'.[39] Montana may well have provided the inspiration for novelist P. G. Wodehouse's 45th book, *Laughing Gas* (1936), a satire of Hollywood and the film industry whose protagonist, Reginald Swithin, 3rd Earl of Havershot, is burdened with a face like a gorilla, 'much more so, indeed, than most gorillas have'.[40]

In the 1960s and '70s another American professional wrestler, Robert Marella, celebrated his enormous weight and strength by adopting the ring name 'Gorilla Monsoon'. He later became a celebrated WrestleMania commentator, announcing matches that were beamed to millions of viewers worldwide.

Gorilla mascot of Pittsburgh State University, Kansas.

Notions of a transfer of gorilla strength and courage to mankind lie behind Dominique Ponchardier (pseud. Antoine Dominique)'s creation in the 1950s of Le Gorille, a French secret agent with supernatural strength and power. Nearly 30 novels as well as successful films featured the immense powers of Le Gorille. The classic film of 1958 starring Lino Ventura, *Le Gorille vous salue bien*, opens with a shot of the Gorille's prison bars bent apart with tremendous force. Later in the film he rips open a massive locked steel door with his bare hands; breaks his chains when temporarily captive; and climbs ape-style up the outside of a multi-storey scaffold. The film's posters captured Le Gorille's dual persona by superimposing silhouetted gorilla mandalas behind the image of Lino Ventura.

The gorilla became symbolic of sporting prowess for the students of Pittsburg State University, Kansas, in 1920, when a group of male students adopted this moniker to rally team spirit for their athletic clubs. By 1925 the animal had become the school's official mascot and remains so today for a wide variety of men's and women's sports. The logo for the World Gym, a global franchise founded by Joe Gold in Los Angeles in 1976, is a gorilla performing a massive power lift. In British and Australian racing slang, 'gorilla' refers to heft of a different calibre, being the term for a £1,000 bet (with a £500 bet referred to as a 'monkey').[41]

The epithet 'gorilla' soon came to have pejorative connotations as well. In 1912 Edward Block, the tallest U.S. government employee in the state of Washington, was written up as 'a human freak' because of the fact that 'the big fellow's gorilla-like arms spread out to a distance of 89 inches.'[42] In early twentieth-century cinema the term 'gorilla' was widely used to denote evil or threatening characters. Gorilla Gulden is a disreputable bandit in Zane Grey's *The Border Legion*, first filmed by Goldwyn in 1918. The thuggish villain in First National's *As Man Desires* of 1925 is called Gorilla Bagsley; while the hired goon in Paramount's *The Fifty-Fifty Girl* of 1928 is Buck, the Gorilla Man. By the 1930s, of course, 'gorilla' had become synonymous with the brute force employed by American gangsters, like the Philadelphia hoodlum murdered outside St Paul in 1932 who went by the name of Harry 'Gorilla Kid' Davis.[43] Gorilla-like goons epitomize the dark side of the Prohibition era. They proliferate in the long-running Belgian comic series *Sammy* (1970–2009), whose adventures are set in 1920s Chicago.

Fun park, Catalonia, 2007.

It is the strong and cruel gorilla police who provide much of the edge to Fox's 'negative utopia drawn from a Darwinian nightmare', *Planet of the Apes* (1968), in which we see them round up, execute and torture helpless humans.[44] A different type of gorilla policing was provided by the Guerrilla Girls, a group of gorilla-mask-wearing feminist art activists who from 1987 onwards picketed the Whitney Biennial and other contemporary art venues and events to protest against the low representation of women artists and against sexism and racism in general.

While earlier Hollywood gorilla suits had been extremely expensive, and their animatronics a topic of occasional press interest, by the 1970s the cheaply manufactured gorilla costume was ubiquitous. At the close of 1969 a Mr Phipps, wearing a rather sad version of the outfit, applied for a librarian's position in Episode 10 of the first series of *Monty Python's Flying Circus*. Across the Atlantic the new decade opened with Carl Reiner's black comedy *Where's Poppa?* (1970), based on Robert Klane's novel. In the film's opening sequences George Segal is shown donning a gorilla suit and then bursting violently in upon his senile elderly mother, Ruth Gordon, in an attempt to scare her to death (a cheaper and easier option than an aged care home). A decade later, the streets of early 1980s Manhattan were to become home to a new tribe of gorilla-suited actors. New York at this time exploded with the gorilla-gram phenomenon, with advertisements offering everything from Yiddish gorillas to stripping gorillas.

Satires on gorilla sexuality became pervasive as the twentieth century progressed. In 1947 Charles Addams drew a classic cartoon for *New Yorker* magazine, 'Oh, oh!', in which a Kong-like ape was caught carrying off a blonde woman by his gorilla wife and children. Two years earlier the exotic dancer Jane Rhodora had enthralled American sailors in Panama, performing her 'Beauty and the Beast' routine as a half-woman half-gorilla. The

Funfair, Jardin des Tuileries, Paris, 2011.

'talk of the fleet', she 'drew mobs' with this act in which the gorilla half of her outfit tore off the rest of her costume.[45] The Swinging Sixties confronted the trope with suitable elan, transporting the lustful gorilla to a Florida nudist camp where he could cavort amidst a bevy of big-breasted nymphettes. While not great cinematic art, *The Beast That Killed Women* (1965) did generate some striking posters featuring drooling gorillas and buxom women. In 1967 gender and sexual politics were both turned on their heads in Ronald Tavel's off-off-Broadway play *Gorilla Queen* in which, in the words of one reviewer of the day, 'the homosexual mafia has now decided to advance the sexual revolution another step'.[46] A send-up of sexual stereotyping in jungle films of the 1930s and '40s, as well as offering a provocatively pre-Stonewall gay homage to the camp intensity of *King Kong*, Tavel's play has been aptly described as forcing 'multiple boundary transgressions: gay and straight, black and white, human and animal, male and female, master and servant'.[47] In an equally camp manner, Friday nights in the 1990s at King Steam, a popular Sydney gay sauna, became reserved for hirsute men and their admirers under the sobriquet of 'Gorillas in the Mist' evenings.

5 All in the Family

Why should our nastiness be the baggage of an apish past and our kindness uniquely human? Why should we not seek continuity with other animals for our 'noble' traits as well?
Stephen Jay Gould, *Ever Since Darwin: Reflections in Natural History* (1977)[1]

In the debates on evolution and creationism that raged throughout the second half of the nineteenth century, the gorilla was lampooned as 'a man and a brother' by the popular press. Darwin himself did not discuss gorillas directly in *On the Origin of Species*, and only in his *The Descent of Man* of 1871 did he address the evolutionary linkages between humans and other primates. It was Darwin's ardent follower Thomas Huxley who wrote openly of humanity's close links with the gorilla, 'the Ape which most nearly approaches man, in the totality of its organisation'. In his *Evidence as to Man's Place in Nature* of 1863, Huxley described the gorilla as 'a brute now so celebrated in prose and verse, that all must have heard of him, and formed some conception of his appearance'. While acknowledging that 'no existing link between Man and the Gorilla' had as yet been proven, Huxley remained convinced that humanity had emerged from 'the gradual modification of a man-like ape'.[2]

Britain's pre-eminent nineteenth-century gorilla specialist, Richard Owen, argued from a position diametrically opposed to that of Huxley. Owen and Huxley had clashed for years over the content and purpose of London's proposed new Natural History Museum, and the contested significance of the gorilla now provided them with a new jousting field.[3] Owen's monograph on the gorilla of 1865 was the first on the subject to appear in the

Joseph Wolf, '*Troglodytes gorilla* mas. adult', 1861, hand-coloured lithograph, from Richard Owen, *Memoir on the Gorilla (Troglodytes gorilla, Savage)* (1865).

English language. While sharing the opinion that the gorilla was in many physical respects the closest of all the apes to mankind, Owen nonetheless reiterated here that 'the genus *Homo* [is] in a distinct group of the mammalian class, zoologically of higher value' than all other primates.[4] For Owen, the fact that the *hippocampus minor* in the gorilla's brain was not as developed as in humans meant that humanity could not possibly be *directly* related to this other primate.

As Huxley and Owen publicly debated issues of evolution and the gorilla, *Punch* magazine entertained its readers with 'Monkeyana', a poem supposedly written in London's Zoological Gardens by a gorilla wearing a sign blazoned with the question 'Am I a Man and a Brother?':

'Monkeyana', cartoon published in *Punch* (18 May 1861).

Am I satyr or man?
Pray tell me who can,
And settle my place in the scale.
A man in ape's shape,
An anthropoid ape,
Or monkey deprived of his tail?

Am I satyr or man?
Pray tell me who can,
And settle my place in the scale.
A man in ape's shape,
An anthropoid ape,
Or monkey deprived of his tail? . . .
Then HUXLEY and OWEN,
With rivalry glowing,
With pen and ink rush to the scratch;
'Tis Brain *versus* Brain,
'Till one of them's slain;
By Jove! It will be a good match![5]

Of crucial importance to everyone interested in evolutionary debate was the second section of Huxley's *Evidence as to Man's Place in Nature*, 'On the Relations of Man to the Lower Animals'. Here he concluded controversially that

> Whatever part of the animal fabric – whatever series of muscles, whatever viscera might be selected for comparison – the result would be the same – the lower Apes and the Gorilla would differ more than the Gorilla and the Man.

The frontispiece to his treatise presented a clear visual comparison between five ambling skeletons – four primate and one human – from London's Museum of the Royal College of Surgeons. This comparative chart, prepared by Waterhouse Hawkins, offered readers a sort of snapshot catwalk of evolution, enabling the leap from gorilla to man to be seen at a glance. The chart has since been imaged endlessly as an icon of evolution in popular culture,

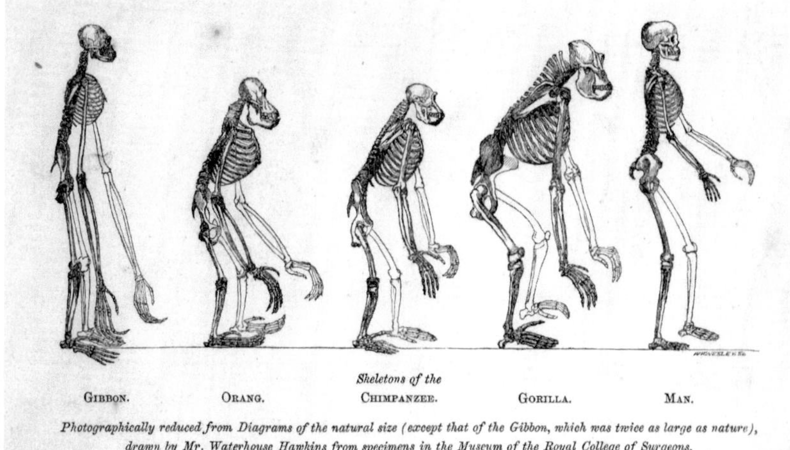

Waterhouse Hawkins, 'Skeletons of the Gibbon, Orang, Chimpanzee, Gorilla, Man', engraved frontispiece to Thomas Henry Huxley, *Evidence as to Man's Place in Nature* (1863).

advertising and art up to the present day. Echoes of Hawkins's comparative chart reverberated almost immediately as far away as Australia when Frederick McCoy, Director of the National Museum of Victoria in Melbourne, placed the skeletons of an adult human female and a male and female gorilla on display in June 1865 alongside his new exhibit, the taxidermied gorilla 'family' that Paul du Chaillu had shot on commission in Gabon. McCoy trumpeted that he had installed this display

> on account of the great scientific interest attaching to the discussions which have arisen between Professor Owen and Professor Huxley, as to whether or not the gorilla formed, in the character of its skeleton, a real link of anatomical or structural passage between man and the lower animals.

He sided firmly with Owen, declaring that through study of Melbourne's new gorilla specimens

'Artistic Jottings at the National Museum', wood engraving, published in *The Australasian Sketcher* (5 August 1876).

> not only the anatomists and general zoologists, but even those interested in the theological question of 'man's place in nature' . . . can satisfy themselves of the impassable gulf in structural sequence which really separates the greatest of the man-like apes from man, viewed merely by the anatomist as a member of the animal kingdom.[6]

The gorilla exhibit at the National Museum of Victoria remained a collection highlight for more than a generation of visitors, as well as the most accessible focal point for local discussion about creationism, Darwin and evolution. Frederick McCoy would doubtless have been pleased to read in *The Australasian Sketcher*, a decade after his acquisition of these preserved specimens, how at the Museum:

> It is not uncommon to see such scenes as our artist has depicted, of groups of fair girls looking at the ruggedly grim family of gorillas and declaring it to be utterly impossible that we could ever be related to those horrid creatures.[7]

In 1863 Thomas Huxley had predicted:

> On all sides I shall hear the cry – 'We are men and women, not a mere better sort of apes, a little longer in the leg, more compact in the foot, and bigger in brain than your brutal Chimpanzees and Gorillas. The power of knowledge – the conscience of good and evil – the pitiful tenderness of human affections, raise us out of all real fellowship with the brutes, however closely they may seem to approximate us.'[8]

In 1872 Charles Darwin's *The Expression of the Emotions in Man and Animals* seemed to bring humanity even closer to the animal kingdom. A highly emotional gorilla, with tears streaming down its face at the thought that it might be the progenitor of the human race, along with Henry Bergh, founder of the American Society for the Prevention of Cruelty to Animals, confronted Darwin with dismay in a witty cartoon published in *Harper's Weekly* in August 1871. Janet Browne has examined how Darwin's uncompromisingly ungroomed features – his great bushy beard and shaggy eyebrows, topped by a prominently round bald pate – lent themselves readily to his transformation into a hybrid gorilla-man at the hands of cartoonists in the 1860s and '70s, creating 'visual statements [which] propelled the idea of evolution out of

MR. BERGH TO THE RESCUE.

THE DEFRAUDED GORILLA. "That *Man* wants to claim my Pedigree. He says he is one of my Descendants."

MR. BERGH. "Now, Mr. DARWIN, how could you insult him so?"

Thomas Nast, 'Mr Bergh to the Rescue', wood engraving, published in *Harper's Weekly* (19 August 1871).

the arcane realms of learned societies into the ordinary world of humor, newspapers, and demotic literature', hastening common acceptance of evolutionism with the gorilla as its centrepiece.[9]

With the rise of evolutionary thinking, ever greater importance was given to attempting to rank humans in the biological order, not only in relation to primates or a speculative 'missing link' between ape and human, but also in relation to each other. The second half of the nineteenth century saw the development of racial pseudo-science as a way of creating a hierarchy of difference within humanity. Human qualities and physical attributes were ranked in order supposedly from least to most evolved. Pseudo-sciences such as craniology, anthropometry and comparative physiognomy were advanced with the purpose of establishing the lesser evolutionary status of non-white peoples. Subsets of humanity were denigrated by the attribution of supposedly ape- or gorilla-like characteristics, actually derived from human characteristics that were considered base (for example, the capacity for acts of violence, particularly murder and sexually motivated attacks).

From the 1860s to the '80s, paralleling the gorilla's increasing visibility in popular culture, cartoonists caricatured as gorillas their political opponents or vilified social groups. In the 1860s members of the Fenian or Irish Nationalist movement were targeted by both English and American satirists as subhuman. Irish rebels were frequently depicted as 'ugly' and 'aggressive' gorillas, with the implication that they were a lesser order within the course of mankind's evolutionary development. Mixing their pejorative visual metaphors for maximum effect, cartoonists of the Victorian era dubbed Irish political agitators as monstrous Frankensteins while also giving them ape-like jut-jawed features or even transposed gorilla heads. Campaigns of racial vilification directed

Matthew Samuel Morgan, 'The Irish Frankenstein', coloured wood engraving, published in *The Tomahawk* (18 December 1869).

Harry Ryle Hopps, *Destroy This Mad Brute: Enlist*, c. 1917, coloured lithograph.

against African Americans in this period also used caricatures that overlaid gorilla-like or simian features in the attempt to associate black people with the 'less evolved' characteristics of non-human primates.[10]

A similar process of demonization occurred during the First World War, when enemy forces were frequently visualized as subhuman. The belligerence of the German military machine was equated with the legendary aggressiveness of the gorilla in Harry Hopps's powerful U.S. recruiting poster, one of the most memorable of its era, which appeared around 1917 at the time of

J. Norman Lynd, *Stop Him!*, 1917, offset lithograph.

America's entry into the First World War. In Hopps's poster the monstrous Prussian gorilla steps onto American soil, brandishing a club in one paw and ravaging the maiden of European civilization with the other. The poster is clearly a reworking of Fremiet's sculpture of the *Gorilla Carrying off a Woman* of 1887. Similar visual rhetoric is found in J. Norman Lynd's *Stop Him!*, a propaganda poster of around 1917 in which the rapacious Prussian gorilla menaces Lady Liberty.

In *Kultur Cartoons* (1915), published in London by Australian artist and satirical cartoonist Will Dyson, German aggression was depicted as a semi-evolved Neanderthal-like gorilla. The original drawings for Dyson's *Kultur Cartoons* were exhibited at London's Leicester Galleries in January 1915, and were also shown in Australia later that year.[11] They were certainly known to Dyson's brother-in-law Norman Lindsay, who throughout the Great War worked for Australia's *Bulletin* magazine, creating allegorical, nationalistic cartoons.[12] In Lindsay's hands, Kaiser Wilhelm and his German troops became personified as the Hun-Gorilla, the beast unleashed upon an innocent world. The Australian public's rejection of conscription at referenda in 1916 and 1917 enraged Lindsay, whose brother Reginald, an infantry gunner, had been killed in 1916. He designed several recruitment posters for the

Norman Lindsay, *Back or Forward?*, c. 1918, pen and ink.

Australian Army, the most powerful of which was his *?* or *Quick!* of around 1918, depicting the marauding Hun-Gorilla stretching blood-drenched fists around the world.

The gorilla was also used throughout the Second World War to demonize social groups and enemy forces. Both Allied and Axis propaganda became decidedly gorilla-centric between 1939 and 1945, each side in the conflict imaging the other's aggression as signifying a pre-evolved primitivism. Gino Boccasile created a poster for Italy's fascist government which depicted a black American GI as a philistine gorilla thief, making off with a Venus de Milo marked down to a paltry $2; while the Russian artists' collective Kukryniksy produced a colour lithograph in 1942 titled *Kill Him!*, showing a Sten gun-toting German gorilla soldier

Norman Lindsay, *?*, or *Quick!*, c. 1916–18, chromolithograph.

Bernard Partridge, after Emmanuel Fremiet, 'John Bull's War Aim', offset lithograph, published in *Punch; or, the London Charivari* (18 October 1939).

trampling on the bodies of women and children. Philip Zec's cartoon 'The New Christianity', published in the *Daily Mirror* in January 1941, showed a gorilla nailing swastika arm extensions onto a cross to satirize Germany's adoption of Nazism as the state orthodoxy; and after the Japanese government executed captured American airmen in 1942, the *New York Times* cartoonist Edwin Marcus imaged Japan as a barbarous gorilla, righteously executed by Roosevelt's United States ('Civilization', represented by a gun aimed at the Japanese gorilla's forehead). Fremiet's *Gorilla Carrying off a Woman* surfaced again on both sides of the conflict, being used in cartoons by Bernard Partridge in *Punch* in October 1939, and also by Oskar Garvens in the German satirical magazine *Kladderadatsch* in November 1941. In Partridge's version Hitler was Fremiet's gorilla, struggling with the maiden Freedom and stalked by a British hunter. Garvens's cartoon, titled 'Roosevelt und die Freiheit' (Roosevelt and Freedom), showed a bespectacled Roosevelt gorilla with the Statue of Liberty trapped in his embrace.[13]

Human prehistory was a common subject in art and literature of the nineteenth and early twentieth century concerned with extemporizing from the evidence of fossil records to produce images of Stone Age ancestors. Gorillas, other primates and visions of 'missing links' often featured. The idea of discovering a 'lost world' with live specimens fired many imaginations. Dinosaurs had been discovered only in the late eighteenth century, along with other prehistoric remains, and the idea of human prehistory dates from this period. The best-known literary visions of prehistory include Sir Arthur Conan Doyle's *The Lost World* (1912) and *The Land that Time Forgot* (1918) by Edgar Rice Burroughs. Notions of 'race', prehistory and 'civilization' also permeated the productions of a youthful Hollywood, reflecting then-dominant paradigms of biological determinism and social Darwinism.

Both Burroughs's and Conan Doyle's 'lost world' novels included anthropoid apes that seemed to be 'throwbacks' from the true path of evolution. In Burroughs's *Tarzan of the Apes* (1912), however, it is civilization that appears unattractive and life in the jungle with the apes far more noble. The figure of Tarzan offered a counterbalance to the perceived problem of emasculation in industrial society at the beginning of the twentieth century and pandered to a taste for wild nature. Tarzan is a natural leader in the jungle, born to nobility as the son of Lord and Lady Greystoke who perished after being marooned on the African coastline. The majority of the other white men in the story prove greedy and violent. The young Lord Greystoke is adopted by Kala, an anthropoid ape, 'a huge, fierce, terrible beast of a species closely allied to the gorilla, yet more intelligent; which, with the strength of their cousin, made her kind the most fearsome of those awe-inspiring progenitors of man'. The apes name him Tarzan, which means 'white skin' in their language. It is after Tarzan has entered his parents' cabin and begun to uncover the mystery of his origins that he encounters a huge gorilla. When it charges him, Tarzan uses the knife he found in the cabin, discovering the effectiveness of man-made weapons. He kills the gorilla, but is gravely wounded himself. Burroughs vividly and imaginatively describes the violence of the gorilla, which, 'fighting after the manner of its kind, struck terrific blows with its open hand, and tore the flesh at the boy's throat and chest with its mighty tusks'.[14]

Paul du Chaillu's accounts of his adventures in the 'Gorilla Country' of Central Africa were a key influence on Burroughs, and incidents described in Du Chaillu's books appear to have been reworked by him for the Tarzan stories. This is also true of Cooper and Schoedsack's gorilla films – *Son of Kong* notably includes a sequence in which Kong Jr breaks a gun, an oft-repeated tale found in the first gorilla accounts of both Savage and Wyman, and Du Chaillu. The best-known cinematic Tarzan stories are the series of MGM features starring Olympic swimming champion Johnny Weissmuller as Tarzan alongside ape-suited actors and chimpanzees, inspired by Burroughs's novel but featuring many newly invented incidents and narratives. The denouement of the first MGM film, *Tarzan of the Apes* (1932), introduces a pit containing a giant gorilla-suited actor (Ray Corrigan, uncredited) with whom Tarzan wrestles to save Jane and her party. While Burroughs and early cinematic adaptations of *Tarzan of the Apes* made a distinction between Tarzan's ape family ('a species closely allied to the gorilla, yet more intelligent') and the ferocious gorilla as foe, over time this has been lost: in Disney's highly successful animated feature *Tarzan* (1999) the ape man's family are gorillas.

Following their runaway success with *King Kong* in 1933, Merian C. Cooper and co-director Ernest B. Schoedsack's later gorilla features, *The Son of Kong* (1933) and *Mighty Joe Young* (1949), show some signs of evolving consciousness with regard to the treatment of large primates. Having created the screen's greatest example of a marauding, womanizing ape (and spawned dozens of subsequent films in the genre), they now sought to humanize the gorilla. *The Son of Kong* was rushed out by RKO Pictures to capitalize on the original's box office success and picks up with the film's central character, director and entrepreneur Carl Denham (Robert Armstrong), bankrupted by damage claims following Kong's rampage through New York City. To escape his creditors, he heads to sea and sails again for Skull Island, this time in search of hidden treasure. The baby gorilla that he now finds on the island is not the symbol of potent blackness his father had been, but a vehicle for exploring Denham's sense of guilt for Kong's death in the first movie. 'Son of Kong' is a cute, light-coloured teddy bear of a gorilla. All of King Kong's sexual charge is neutralized, and the big baby seems more like a surrogate child or pet for Denham and his new love interest Helen. Notable is a sequence in which the Son of Kong's wounded finger is bandaged with parental care by Denham. *King Kong*'s rampaging gorilla featured some ambiguous facial expressions, but was very limited in his emotional range. *Son of Kong* constructs baby Kong as a 'good gorilla' by comically exaggerating his human qualities – his ability to seek sympathy when he feels pain, his

Jacques de Loustal, *Tarzan Saving Jane*, 2009, ink and watercolour.

head-scratching thinking gestures and his appreciation of romantic innuendo. Perhaps inspired by *Son of Kong*, Tintin meets a gorilla in the denouement of *L'Île noire* (*The Black Island)*, first published in 1937 in *Le Petit Vingtième*. This initially fearsome beast turns out to be a big baby that cries when it falls down a flight of stairs, and afterwards sports a bandage on its wrist.

In the last of Cooper and Schoedsack's big ape trilogy, *Mighty Joe Young* (1949), Joe the gorilla is raised in Africa by an American girl, Jill, then taken to Los Angeles by the promoter Max O'Hara to perform in his nightclub, the 'Golden Safari'. Joe, wrongfully plied with liquor, goes on the rampage but redeems himself by helping rescue infants from a burning orphanage. The remorseful O'Hara helps Jill and her new husband to return Joe to Africa, which, as sympathetic colonials in tune with the gorillas, is the newlywed couple's natural home. Like Tarzan, Jill and Joe belong outside of the taint of civilization.

The notion of gorillas providing an alternative family environment, morally superior to that found in human society has been frequently explored in both literature and film, often with echoes of Burroughs's original *Tarzan of the Apes*. In Warner Brothers's *Gorilla My Dreams* (1947), Bugs Bunny washes ashore on an island inhabited by talking, 'Apes of Wrath'-reading primates, where he is adopted as a foundling by the childless Mrs Gruesome Gorilla. This is much to the chagrin of her husband, Gruesome, who seeks for the remainder of the cartoon to eradicate Bugs from the new family equation, much as the ape husband of Tarzan's adoptive mother Kala attempts to get rid of the human child interloper. Similarly, John Turteltaub's *Instinct* (1999) showed Anthony Hopkins abandoning his own family in order to be adopted by a gorilla group in Africa, with tragic consequences for all concerned. The film offered a very loose riff on Daniel Quinn's metaphysical novel *Ishmael* (1992) about a telepathic gorilla that engages in Socratic dialogue on human folly, its roots and remedies. The ultimate cinematic expression of the idealized gorilla family remains *Gorillas in the Mist* (1988), which, it has been argued, 'decenters humanity in the world by repositioning it within an animal context'. Selectively ignoring Dian Fossey's own accounts of gorilla violence and discord, the film presents the species as 'natural, pure, clean, loving, non-violent and innocent. They never once do anything cruel to each other or to humans. People, on the other hand, are depicted as destructive, violent, selfish, egotistical, dirty and disdainful, or at best oblivious to nature'.[15] An inversion and hybridization of the happy gorilla family trope is explored in Maureen Duffy's *Gor Saga* (1981) and its subsequent adaptation as the BBC's miniseries *First Born* (1988). Here a female gorilla, which has been impregnated by a genetic researcher with his own sperm, brings the resulting hybrid baby successfully to term. While betraying his human wife and daughter, the scientist also fails to provide a successful home life for his illegitimate half-gorilla son, who would have been better off, it seems, in the wild with his mother.

Many early horror films featured cross-species experimentation with gorillas and other apes. In the nineteenth and early twentieth centuries, it was widely believed that blood, spinal fluid or bodily organs could convey intellectual and emotional qualities when transfused or transplanted. These evolutionary and scientific obsessions metamorphosed into cinema's legendary monsters of the 1930s, '40s and '50s. Cross-breeding is a central leitmotif in Erle C. Kenton's *Island of Lost Souls* (1932), in which the demented Dr Moreau (Charles Laughton) has created a new servile class from, among other variants, human/gorilla experiments. As well as assisting makeup supervisor Wally Westmore with making furry chest and back pelts for the dozens of actors portraying Dr Moreau's male mutants, Charlie Gemora also appeared briefly here as a lively gorilla.[16] In 20th Century Fox's *Dr Renault's Secret* (1942), Renault's apish handyman Noël turns out to be a gorilla who has been humanized by the mad doctor's radical treatments. Noël, we learn as the film unfolds, is the product of 'experimental humanization' involving glandular injections and brain surgery, followed by plastic surgery to give him the passable features of a human being. The transplantation of still-living brains from the corpses of murdered men into new, vengeful gorilla bodies informed both Don Wilcox's novella *The Whispering Gorilla*, published in *Fantastic Adventures* in May 1940, and Paramount's gangster film *The Monster and the Girl* (1941).

In Robert Florey's adaptation of Edgar Allen Poe's *Murders in the Rue Morgue* (1932), a film deeply imbued with Darwinian

Cover of *Fantastic Adventures* (May 1940).

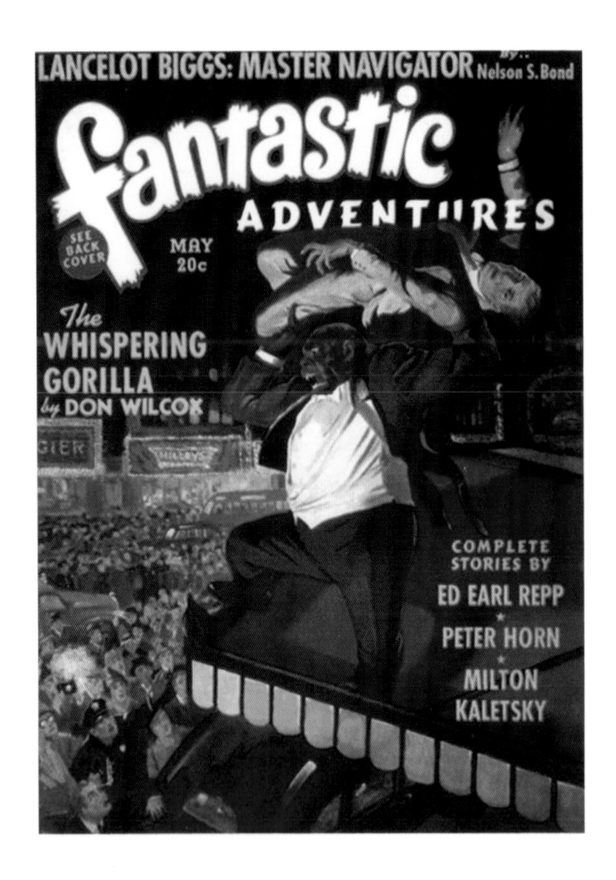

theory, Erik the gorilla (played, uncredited, by Charles Gemora) is the 'creature' of the sinister Dr Mirakle (Bela Lugosi), who extols the close relationship that exists between his gorilla and mankind, citing Erik's capacity for experiencing emotions as evidence of his quasi-human status and introducing him as 'the darkness at the dawn of man'. Mirakle claims that Erik, 'the Beast with a human soul', is not a freak but a 'milestone in the development of life'. When Mirakle delivers a disquisition on evolutionary theory ('The shadow of Erik the Ape hangs over us all'), we see on the wall behind him Huxley's famous 'catwalk' line-up of primate and human skeletons. When Mirakle and his assistant 'Janos, the Black One' (Noble Johnson, who also plays Skull Island's native chief in *King Kong*) unveil Erik as 'the first man', the audience is scandalized by this heresy. Mirakle promises to prove his contentions through scientific experiments which involve the transfusion of ape blood into young women. In Arthur Conan Doyle's late Sherlock Holmes tale 'The Creeping Man' (1923), the simian antics of the famous physiologist Professor Presbury have also been brought about by his injecting himself with serum derived from the black-faced Himalayan Langur monkey.

It was not only in film and fiction that this sort of 'mad science' was in the air. In the 1920s the French/Russian surgeon Serge Voronoff became famous (and later infamous) for his experimental surgery, grafting slices from the testes of chimpanzees and baboons into the testicles of human patients as a means of rejuvenating elderly men. The rush on his services was so great, he told the *Chicago Tribune* in 1922, that 'only the scarcity of chimpanzees has prevented him from performing daily operations in his Paris laboratory, as he has a long waiting list. He is using chimpanzees faster than they can be caught in Africa.'[17] Voronoff was soon dreaming of monkey farms established in Europe that could 'constitute great factories designed to supply spare parts for the human machine'.[18] In 1931 the biologist Constantine Leventis privately published a treatise, *Sex Glands Function and the Human Life*, which advocated 'grafting the gonads of the civilized human organism onto the organism of chimpanzees or gorillas' in order 'to obtain a species nearer to the human race, if not quite resembling it . . . a missing link between the human race and the higher apes'.[19] In the late 1920s the Russian artificial insemination expert Il'ya Ivanov had experimented with impregnating female chimpanzees with human sperm, and also envisaged inseminating human females with orang-utan sperm, in order to produce a hybrid species that would provide 'a better understanding of the problem of the origin of man'.[20] Bertram Gayton explored the extreme dream in his novel *The Gland Stealers* (1933). Here a nonagenarian, inspired by Voronoff's gland-grafting rejuvenation experiments, decides to acquire a gorilla for the ultimate in transplanted virility. Following his post-operative restoration of libido, he takes a hundred elderly men to Africa in search of their own gorilla glands, where it soon becomes evident that demand greatly outstrips supply.[21]

Film noir legend Edward Dmytryk brought directorial flare to *Captive Wild Woman* of 1943, a Hollywood potboiler about glandular transplants between gorilla and human subjects. Aided by an involuntary donation from his theatre nurse (whom he murdered in order to advance his experiments, extracting her cerebrum), Dr Sigmund Walters (John Carradine) transforms Cheela the gorilla on the operating table into 'a human form with animal instincts'. He releases her into the world as Paula Dupree (Acquanetta), a mysterious dark-skinned beauty whose uncanny ability to tame wild beasts with her stare makes her an integral part of the circus lion-and-tiger act of Fred Mason (Milburn Stone), the animal trainer who had originally brought Cheela from Africa to the U.S. Paula's composure lasts until her jealousy of Mason's girlfriend causes her to regress into animal form. In a climactic scene Paula/Acquanetta sees herself transform before a mirror firstly into a black woman, and then into a gorilla. Two further 'Gorilla Girl' films were completed in the following years: *Jungle Woman* (1944) – in which it is revealed that Paula began life as an African woman and was turned into a gorilla by a witch doctor – and *The Jungle Captive* (1945).

Zamora the Gorilla Girl, Ohio State Fair, 1999.

The transformation from girl to gorilla in *Captive Wild Woman* quickly crossed into popular culture and became the theme for a popular American sideshow performance known as Zamora the Gorilla Girl. Patented by A. K. Brill of Peoria, Illinois, who began building carnival games in 1941, Zamora terrified generations of children at theme parks across the U.S. as, under mood lighting, an African American woman imprisoned in a cage appeared to metamorphose before the viewers' eyes into a growling ape. The climax of the act came when the newly formed female gorilla rushed the bars of her cage, which dropped to unleash the creature directly into the midst of a delightedly screaming audience. This 'Girl to Gorilla' sideshow act was immortalized in the James Bond blockbuster *Diamonds are Forever* (1971).[22]

The preoccupation with blood and transplantation in the 1930s and '40s developed into an interest in the powers of radiation and in genetics in the 1950s. *King Kong vs. Godzilla* (1962) stages a clash between a throwback of evolution, the giant ape, and the giant mutated lizard, a monster of the atomic age. The story follows a similar trajectory to the original *King Kong*: adventurers journey to a remote tropical island where they discover and capture the giant Kong. Kong escapes and climbs the tallest building in Tokyo, a screaming maiden in hand. Meanwhile, on the other side of the world, Godzilla breaks out of an iceberg and begins his return journey to Japan, leading to a fateful meeting of the two monsters. Willis O'Brien, the master of stop-motion animation who brought the King Kong of 1933 to life, initially conceived of

Silverback gorilla from the 'Great Ape' exhibit at Denver Zoo.

the idea of pitting King Kong against Frankenstein's monster, but the giant gorilla's rival monster was changed into Godzilla when the concept was sold to the Japanese film studio Toho.

Science fiction and horror film visions of gorilla experiments were reversed in *Planet of the Apes*'s cinematic commentary on primate experimentation of 1968, based on Pierre Boulle's novel of 1963. Caged humans captured by violent gorilla troops are subjected to behavioural research and surgical intervention by chimpanzee scientists under the direction of conservative orang-utan leaders. The humans cannot speak and the apes do not believe that they have higher intelligence, so feel free to subject them to invasive experimentation. A parallel reversal had previously been explored in DC Comics's Strange Adventures series, issue #45, 'The Gorilla World!' (June 1954). The cover features a bare-chested man bolted to an upright restraining table, with the ape doctor who is examining him saying 'Incredible! This puny creature is the *missing link* between ape and animal!'[23]

Historically, the research of evolutionary theorists and primatologists has often been directed in such a way as to naturalize particular human behaviours by showing that these have an evolutionary function for gorillas and other apes, as close relatives of human beings. Scientists have assumed that aggression, competition and male dominance characterize humanity in the 'natural state' and have looked for these behaviours to be naturally selected amongst primates – domination and hierarchy having been widely postulated as bases for social progress, rather than cooperation or empathy. On the other hand, higher characteristics often considered as defining a break between humans and other great apes – notably empathy and altruism – have been assumed to be outside of evolutionary forces and natural selection. In his essay 'Primates and Philosophers: How Morality Evolved' (2006), primatologist Frans de Waal discusses the vexed question of morality and nature in the history of evolutionary theory, with Darwin in *The Descent of Man* (1871) clearly treating an ethical sensibility as an integral part of human nature where his follower Huxley considered it a victory over evolution and nature. De Waal maintains that ethical qualities and behaviours exist in other animals and that we evolved from non-human ancestors that cooperated and cared for others, cooperative rather than selfish behaviour being naturally selected. He sees the evolutionary origin of an ethical tendency to act for common good in emotions that humans share with great apes, notably empathy, rather than in an exclusively human morality with a rational basis.

This runs counter to research frameworks that have assumed that higher capacities and qualities are unique to human beings. It was first postulated that primates were incapable of conceiving existing objects as tools or of fashioning tools themselves, and that this ability set humans apart from them. In his early field research on mountain gorillas, for example, Schaller recorded no tool-using and attributed this to the abundance of easily harvested vegetation in the animal's habitat, concluding that 'no preparation of the food is required beyond stripping or shredding which is readily accomplished with the teeth and fingers.'[24] Yet he recorded their careful handling of prickly plants, folding them for sting-free eating.[25] Chimpanzees were observed in the wild in Liberia as early as 1951 using rocks to open nuts, and subsequently using sticks and other objects as tools, primarily for food sourcing.[26] It was only in 2005 that tool use was finally also recorded among gorillas in the wild. One regularly observed female Western lowland gorilla, known as Leah, used a metre-long stick to test the depth of a large pool of water she was attempting to wade across in the Mbeli Bai region of the Nouabalé-Ndoki National Park in the Republic of Congo. Another female gorilla, named Efi, held a stick for support while she foraged for food with the other hand, and then used the stick as a bridge to cross over swampy ground.[27] These actions, while seemingly simple in themselves, suggest cognitive and problem-solving abilities that are shared with humanity. To Gillian Sebestyen-Forrester, who has researched gorilla communication at the University of Sussex, 'this suggests that the gorilla is capable of some mental calculation and abstract thought'.[28] Prior to this observation only gorillas in captivity had been known to use tools. The overturning of accepted knowledge on this subject shows how little we still really know about the gorilla, despite decades of field observation. Leah was also observed by astonished researchers in 2008 mating with George, the silverback in her family group, the pair coupling face-to-face and looking into one another's eyes – previously assumed to be uniquely human sexual behaviour as an indicator of the enjoyment of sex independent of the outcome of procreation.[29]

From the publication of Charles Darwin's *The Expression of the Emotions in Man and Animals* in 1872 and before, writers as well as scientists had speculated upon the emotional and communicative capacities of animals. The rampaging orang-utan in Edgar Allan Poe's *The Murders in the Rue Morgue* of 1841, for example, experiences pleasure, anger, murderous wrath and fear during the course of the tale. Darwin's publication lent scientific credence to such notions, suggesting notably that a range of facial expressions are shared among higher species to communicate emotional states. In 2008 the world's press was fascinated by the plight of the female gorilla Gana in Münster Zoo in northern Germany when confronted with the death of her infant son Claudio. One reporter described how

> She carried his corpse everywhere, guarding the little figure so zealously that wardens at the zoo were unable to retrieve the dead baby gorilla for four days. But it wasn't

> her anger that left a lasting impression; it was the rawness of a mother's pain. So clearly inconsolable, bewildered and shattered, her face displayed a range of emotions that we once thought of as uniquely human. This week she was a grieving mother first, an ape second . . . The DNA of gorillas and humans is 99.9% identical, yet it is one thing to accept in theory the extent to which we are related to the great apes, quite another to witness it.[30]

The need for apes to mourn is being recognized in the zoo world, where it is becoming an accepted practice to allow family members to see the body and say goodbye, since when an animal disappears without trace its companions react badly. Behaviours testifying to higher emotions such as empathy have also been observed, as for example when in 1996, at Chicago's Brookfield Zoo, a female called Binti Jua was sitting with her gorilla family when a toddler fell into the gorilla enclosure. Binti Jua, who had an infant clinging to her body, walked over, picked up the human infant and carried him over safely to the zoo staff. A similar event had occurred at Jersey Zoo in 1986 when the male gorilla Jambo defended an unconscious boy who had also fallen into the gorilla enclosure, stroking him empathetically.[31]

One of the areas of research with chimpanzees and gorillas to attract the most widespread interest and attention has been that of language. Gorilla vocalizations were the focus of research from the earliest observations of gorillas in the wild. Richard Garner notably undertook studies in this field as early as 1890, arguing that vocalizations should be considered as forming a language.[32] Although the non-delivery of his phonograph prevented him from making live recordings of gorilla sounds during his field study in the Congo in 1893, Garner returned from that trip declaring that 'the natives, like myself, are firmly convinced that

Marco Stepniak, *Gana Mourning Her Dead Son Claudio, Munster Zoo, September 2008.*

the gorilla and the chimpanzee tribes have a language.'[33] Despite his lack of recording equipment, Garner noted how through transcription 'I recorded about twelve words of the chimpanzee language, and about half as many of those of the gorilla tongue. I found that the two were entirely distinct.'[34] Studies since that time have confirmed that gorillas use defined vocalizations – grunts, growls, roars, whines, barks and other noises – to communicate in combination with gestures, postures and behaviours including chest-beating, staring or avoiding eye contact, charging and running sideways, sticking out their tongues and moving their lips, hand clapping and throwing things.[35] In primate research in the U.S. after the Second World War a strong interest emerged in interspecies communication. This reflected a broader popular and academic interest at the time in language acquisition, influenced by linguistic theories, such as those of Noam Chomsky, postulating the existence of innate human syntactical structures. The continuing testing of evolutionary frameworks inevitably led to the question of the existence of related language ability in higher primates. Over the period from 1947 to 1953 Kathy and Keith Hayes attempted to teach the chimpanzee Vicki spoken English but declared this a failure. In his primate studies from the 1920s Robert Yerkes noted the ability of chimpanzees to understand human speech but not to reproduce it, and the ability to copy human actions; other researchers subsequently recorded the use of meaningful gestures by gorillas – as well as chimpanzees, bonobos and orang-utans – to communicate among themselves. It is now recognized that apes do not possess the necessary vocal apparatus for spoken human languages, and that, rather than comparison of human speech and animal vocalizations, gestures may hold the key to better understanding the evolution of language.[36]

The National Geographic Society funded successful initiatives in the 1960s to teach American Sign Language (AMESLAN) to chimpanzees. From 1966 Allen and Beatrice Gardner taught the young female chimp Washoe to sign and Roger Fouts continued this work with Washoe and other chimps. Soon afterwards, from 1972, Penny Patterson, a graduate student in psychology from Stanford University, began to teach sign language to two Western lowland gorillas – Koko, born at San Francisco Zoo in 1971, and later Michael, born in Cameroon in 1973, who arrived in the U.S. at the age of three. In 1976 Patterson established the Gorilla Foundation, a research endeavour described on its website (koko.org) as the longest-running ongoing interspecies communication and education project. Patterson has recorded how within two weeks of starting to learn sign language Koko was correctly performing initial gestures including those for food and drink and that this has developed to a signed vocabulary of over 1,000 words and an even greater understanding of spoken English. Michael had learned over 500 signs by the time he died of heart failure in 2000 (cardiovascular disease is a common cause of death in adult male gorillas in captivity). The Gorilla Foundation's blog details many incidents demonstrating that Koko has a sense of humour, feels empathy, remembers past events and recognizes herself in a mirror – all capacities or behaviours widely considered unique to

Koko, Western lowland gorilla, with her pet kitten. Image courtesy The Gorilla Foundation/ Koko.org.

humanity. The website also recounts her extended mourning behaviour after Michael's death. Michael is described as having been a gifted gorilla who 'spent his days signing, painting, listening to and making music and creating vivid works of representational and abstract art'. Patterson further recounts that he could tell stories of past events in sign and had described the murder of his family by poachers.[37]

When still a young gorilla, Koko asked for a cat and was allowed to choose a pet kitten from a litter. She selected a male with no tail and named it 'All Ball'. Koko's gentle care of the kitten, and of other pet dogs and cats after All Ball died in 1984, led researchers to hope that this was 'surrogate mother' behaviour and would lead to starting a family, and to the possibility of observing her teaching signs to her offspring, but this has not yet happened. Koko appeared on the cover of *National Geographic* magazine in October 1978 with a camera, photographing herself in a mirror. Patterson recounts that when Koko saw the photograph she recognized herself and signed 'love camera'. In January 1985 she again appeared on the *National Geographic* cover, this time with her pet kitten. The two images crystallize questions of similarity to humans, with Koko recognizing herself in mirrors and keeping pets like a human. The American Society of Magazine Editors chose Koko's self-portrait as one of the 40 best magazine covers of the last 40 years in 2005.[38]

The Koko phenomenon crossed readily over into mainstream cultural debate. It was parodied in the British television satire *Not the Nine O'Clock News* in an episode that aired in April 1980. This contained a TV current affairs 'live interview' with Professor Timothy Fielding and Gerald the gorilla, whom the anthropologist had allegedly captured in the Congo in 1968 and taught to speak. As Gerald, with an Oxbridge accent, corrects his captor's diction, teaching methods and general gorilla knowledge, the anthropologist is reduced to spitting back: 'You arrogant little bastard, you're wrecking my life's work!' In Paramount's *Congo* (1995), starring Laura Linney and Tim Curry, based on Michael Crichton's science fiction novel of 1980 about a King Solomon's Mines-like search for superconductor diamonds in a lost city in the Congo jungle, Amy the talking gorilla is equipped with a computerized backpack which translates her sign language into human speech, synthesizing an infantile squeaking voice. A more believable and sensitive treatment of the 'intelligent' gorilla was seen in 'Fearful Symmetry', from the second season of the television series *The X-Files*, which aired in the U.S. in February 1995. Here a lowland gorilla named Sophie, who has been taught to sign some 1,000 words, is a key witness to the activities of alien conservationists who have been impregnating endangered animals in an Idaho zoo.

Barbet Schroeder's documentary *Koko: A Talking Gorilla*, which screened in the Un Certain Regard section at the 1978 Cannes Film Festival, shows Koko and Penny Patterson together in teaching interactions and daily activities. The film begins with the six-year-old Koko looking into a ViewMaster at images of a bird, and ends with her looking straight into the lens at the viewer. Schroeder became impassioned about the plight of the gorilla at the time he made the film, and went to Africa hoping to visit Dian Fossey, but when he wrote to her she replied that she would shoot him if he came near her camp, a position that he says was understandable given the ongoing devastation wrought on the gorilla and its habitats.[39] The concluding sequence of Schroeder's film features a discussion with Patterson of Koko's status as a 'person' with rights, suggesting the need for a new understanding of what it is to be a person.[40] In a *National Geographic* article published in the same year as the film's release entitled 'Conversations with a gorilla', Patterson recounts the following interchange: 'I turn to Koko:

Gorilla-suited Steve Calvert with Raymond Burr in *Bride of the Gorilla* (1951). Courtesy Wade Williams Distribution.

"Are you an animal or a person?" Koko's instant response: "Fine animal gorilla."'[41] Self-consciousness of this sort has often been considered to be unique to humanity.

A broad recognition of the proximity of humans to other animals has arisen over recent decades in the light of popular reporting of findings such as the human genome project's confirmation at the turn of the millennium that 90 per cent of human cells derive from the same genetic material as the simplest forms of animal life. The Great Ape Project, founded in 1993 by philosophers Peter Singer and Paola Cavalieri, aims to secure expanded rights for the non-human higher primates – chimpanzees, bonobos, orang-utans and gorillas. Its advocacy underlines the genetic and behavioural similarities between humans and other hominids that also display emotions such as fear, and should therefore be protected by similar laws. In 2008, inspired by this initiative, the government of Spain passed a resolution promising basic 'human' rights to higher primates and ensuring a ban on scientific experimentation involving higher-level primates, as well as on using apes in circuses, television commercials or filming. The bill did not ban zoo exhibition but aimed to greatly improve zoo living conditions. Countries that have banned animal experimentation on gorillas, chimpanzees, bonobos and orang-utans include Britain, Australia and New Zealand.[42] The gorilla's physical resemblance to human beings, as well as its plight at the hands of humanity, places it at the centre of contemporary philosophical and ethical debates which call into question a long history of human exceptionalism, whereby humans have considered themselves to be unique within – or set apart from – the natural world.[43] Moreover, all that has been denigrated as uncontrolled and animal in human nature has been projected particularly onto the gorilla, creating a monster of repressed animality. In order for the gorilla to survive, rapid transformation must be effected in assumptions that the natural world exists for the benefit of humanity, transformation grounded in a thoroughgoing understanding of the interdependent animal nature of the human.

Timeline of the Gorilla

c. 10 MYA

The lineage of the gorilla species branches away from that of humans and chimpanzees, referred to as the gorilla speciation event

5TH CENTURY BC

Hanno II of Carthage encounters the 'gorillae' during a voyage around the western coast of Africa

1560S

Andrew Battell learns of 'two kinds of Monsters' in the Congo – the *Pongo* and the *Engeco*

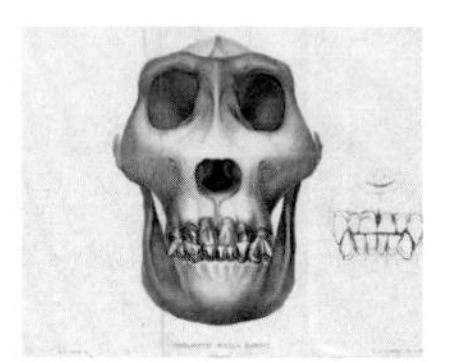

1847

Thomas Staughton Savage and Jeffries Wyman first publish a scientific description of the gorilla in the *Boston Journal of Natural History*

1887

Emmanuel Fremiet unveils a second life-size sculpture, *Gorilla Carrying off a Woman*, at the Paris Salon

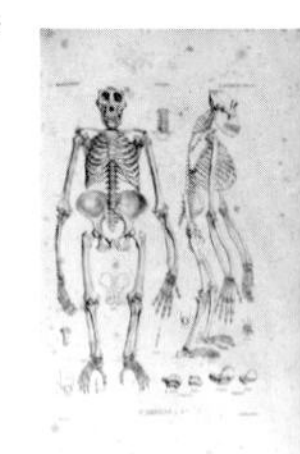

1902

Robert von Beringe comes across the mountain gorilla in the Virunga Volcanoes. His colleague in Berlin, Paul Matschie, names it *Gorilla beringei*

1910

French colonial administration in Gabon prohibits Fang *Ngi* gorilla cults

1972

Penny Patterson begins to teach sign language to Koko, a Western lowland gorilla

1983

Dian Fossey publishes *Gorillas in the Mist*

1993

The Great Ape Project is founded by philosophers Peter Singer and Paola Cavalieri with the aim to secure expanded rights for the non-human higher primates: chimpanzees, bonobos, orang-utans and gorillas

2005

Tool use is observed among gorillas in the wild

1849

The first complete gorilla specimen is transported from Gabon to France, and donated to the Muséum d'Histoire Naturelle in Paris

1859

Charles Darwin publishes *On the Origin of Species*; Emmanuel Fremiet exhibits his gorilla sculpture, *Gorille femelle*, at the Paris Salon

1860s

The 'Gorilla Wars' break out in London, Melbourne and other cities

1861

Paul du Chaillu publishes *Explorations and Adventures in Equatorial Africa*

1926–7

The necrophiliac 'Gorilla Man' murderer terrorizes the U.S. and Canada

1933

Merian C. Cooper and Ernest B. Schoedsack produce and direct RKO's *King Kong*

1936

Carl Akeley's gorilla diorama completed in the African Hall at New York's American Museum of Natural History

1954

Twentieth Century Fox's 3D extravaganza, *Gorilla At Large*, is released

1956

The first gorilla is born in captivity at the Columbus Zoo in Ohio

2008

The government of Spain passes a resolution promising basic 'human' rights to higher primates and ensuring a ban on scientific experimentation involving higher-level primates

2009

The United Nations proclaims the Year of the Gorilla

2012

The genome sequence of a gorilla is assembled the first time with data drawn mainly from a female Western lowland gorilla, making it possible to compare the DNA of the four great apes, human, chimpanzee, gorilla and orang-utan

References

1 LASCIVIOUS BEAST OR SHY VEGETARIAN?

1 Richard L. Garner, *Gorillas and Chimpanzees* (London, 1896), p. 214.
2 Donna Haraway, *Primate Visions: Gender, Race, and Nature in the World of Modern Science* (New York and London, 1989), pp. 2, 5, 10–13.
3 See Christian Nellemann, Ian Redmond and Johannes Refisch, eds, *The Last Stand of the Gorilla: Environmental Crime and Conflict in the Congo Basin, A Rapid Response Assessment* (United Nations Environment Programme, GRID-Arendal, 2010); and 'Mountain Gorilla Census Results: Population Increases by 26.3%', www.gorillacd.org, accessed 17 March 2011.
4 Galina V. Glazko and Masatoshi Nei, 'Estimation of Divergence Times for Major Lineages of Primate Species', *Molecular Biology and Evolution*, XX/3 (2003), pp. 424–34, available at mbe.oxfordjournals.org, accessed 20 March 2011.
5 See Arthur J. Riopelle, '"Snowflake": The World's First White Gorilla', *National Geographic*, CXXXI/3 (March 1967), pp. 443–8.
6 Alan F. Dixson, *The Natural History of the Gorilla* (London, 1981), p. 46.
7 For an interesting summary of gorilla sexuality, see Geoffrey H. Bourne and Maury Cohen, *The Gentle Giants: The Gorilla Story* (New York, 1975), pp. 243–62. See also David P. Watts, 'Mountain Gorilla Reproduction and Sexual Behaviour', *American Journal of Primatology*, XXIV/3–4 (1991), pp. 211–25.

8 'Zoos often classify all specimens from the eastern part of the gorilla's range as "mountain" gorillas (*G. g. beringei*), whereas most of these in fact belong to *G. g. graueri*, the eastern lowland subspecies'. Dixson, *The Natural History of the Gorilla*, p. 9.
9 Stephan Koja, ed., *Gustav Klimt: The Beethoven Frieze and the Controversy over the Freedom of Art* (Munich, 2007), p. 101.
10 'An Immense Gorilla: Monster of Immense Proportions Recently Killed in Colony of Cameroon', *The Washington Bee* (5 June 1901); the article is accompanied by an engraving of Paschen's dead gorilla.
11 Percy Home, 'The Shooting of the Hugest Known Gorilla', *The Sphere* (20 July 1901), p. 68.
12 'Biggest Gorilla Ever Seen By Man', *Los Angeles Times* (17 August 1901).
13 Tamara Giles-Vernick and Stephanie Karin Rupp, 'Visions of Apes, Reflections on Change: Telling Tales of Great Apes in Equatorial Africa', *African Studies Review*, XLIX/1 (April 2006), p. 59.
14 Louis Jacolliot, *Voyage au pays des singes* (Paris, 1883), pp. 119–20.
15 Ibid., pp. 123–4.
16 Paul B. du Chaillu, *Explorations and Adventures in Equatorial Africa* (New York, 1861), p. 87.
17 Wilfred P. Pond, 'Hunting the Gorilla', *The Anaconda Standard* (6 December 1889).
18 See Giles-Vernick and Rupp, 'Visions of Apes', pp. 63–6. Pierre Ichac's 'Le Gorille et le vieux nègre' ('The Gorilla and the Old Black Man'), which appeared in the *Journal des voyages et aventures* in August 1948, purported to tell of the author's recent experiences in Cameroon, where he said local forest dwellers believed that the spirits of sympathetic gorillas and elderly men sometimes fused inseparably, such that when one died so did the other. Pierre Ichac, 'Le Gorille et le vieux nègre', *Journal des voyages et aventures de terre, de mer et de l'air*, LXXII/121 (August 1948), pp. 2279–80.
19 Philippe Laburthe-Tolra, *Les Seigneurs de la forêt: Essai sur le passé historique, l'organisation sociale et les normes ethiques des anciens Bëti du Cameroun* (Paris, 1981), p. 427; cited in Giles-Vernick and

Rupp, 'Visions of Apes', pp. 56–7; Robert H. Nassau, *Where Animals Talk: West African Folk Lore Tales* (London, 1914).

20 Du Chaillu, *Explorations and Adventures in Equatorial Africa*, pp. 306, 352.

21 W. Winwood Reade, *Savage Africa: Being The Narrative of a Tour in Equatorial, South-Western, and North-Western Africa: With Notes on the Habits of the Gorilla, on the Existence of Unicorns and Tailed Men: On the Slave Trade: On the Origin, Character, and Capabilities of the Negro and on the Future Civilization of Western Africa*, 2nd edn (London, 1864), vol. I, pp. 194–5.

22 Günter Tessmann, *Die Pangwe: Volkerkundliche Monographic eines westafrikanischen Negerstammes* (Berlin, 1913); Alexandre Pierre, 'Proto-histoire du groupe beti-buku-fang: Essai de synthèse provisoire', *Cahiers d'études africaines*, v/20 (1965), pp. 503–60.

23 James W. Fernandez, *Bwiti: An Ethnography of the Religious Imagination in Africa* (Princeton, NJ, 1982), pp. 99–100, 116–17, 394.

24 Angela Meder, 'Gorillas in African Culture and Medicine', *Gorilla Journal* (18 June 1999), www.berggorilla.org; Dale Peterson, 'Great Apes as Food', *Gastronomics: The Journal of Food and Culture*, III/2 (Spring 2003), p. 65.

25 Du Chaillu, *Explorations and Adventures in Equatorial Africa*, p. 101.

26 Ivan T. Sanderson, 'An Expedition to the British Cameroons', *The Geographical Journal*, LXXXV/2 (February 1935), p. 130.

27 Giles-Vernick and Rupp, 'Visions of Apes', p. 55.

28 Meder, 'Gorillas in African Culture and Medicine'.

29 See for example Jorge Sabater Pi and Colin Groves, 'The Importance of the Higher Primates in the Diet of the Fang of Rio Muni', *Man*, VII/2 (June 1972), pp. 239–43; Evan Bowen-Jones and Stephanie Pendry, 'The Threat to Primates and Other Mammals from the Bushmeat Trade in Africa, and How This Threat Could Be Diminished', *Oryx*, XXXIII/3 (1999), pp. 233–46.

30 Conrad Gesner, *Historiae animalium* (1551). Janson notes that the illustration of a strange lion-headed, pendulous-breasted and baboon-bottomed *Homo sylvestris* that appeared in Gesner's publication was itself based upon an even earlier source,

a woodcut depicting the exotic animals of the Near East in Bernhard von Breydenbach's *Travels*, published in Mayence in 1486. H. W. Janson, *Apes and Ape Lore* (London, 1952), pp. 332–4.

31 Samuel Purchas, *Haklvytvs posthumus; or, Pvrchas his Pilgrimes: Contayning a history of the world, in sea voyages, & lande-trauells, by Englishmen and others . . . Some left written by Mr. Hakluyt at his death, more since added, his also perused, & perfected. All examined, abreuiated, illustrates w[i]th notes, enlarged w[i]th discourses, adorned w[i]th pictures, and expressed in mapps. In fower parts, each containing fiue bookes* (London, 1626), vol. V, p. 623.

32 Georges-Louis Leclerc, Comte de Buffon, *Histoire naturelle, générale et particulière, avec la description du Cabinet du Roi* (Paris, 1749–66), vol. XIV p. 30. It took further decades of research, down until the opening of the nineteenth century, for European natural historians like the Comte de Buffon, Petrus Camper and their followers to establish a distinction between the chimpanzee and orang-utan.

33 Leslie Lieber, '"I'm Tired Of Being An Ape"', *Los Angeles Times* (9 February 1947).

34 See *Hollywood Goes Ape!* (1994), recently made available on the compilation DVD *Dinosaurs vs. Apes*, released by Frontline Entertainment and Cinema Epoch in 2007. An excellent and comprehensive website devoted to gorilla men in the movies is run by Gorilla Man at www.hollywoodgorillamen.com, accessed 20 October 2012.

35 Groucho Marx, *Groucho and Me* [1959] (London, 2002), p. 176. On gorillas as comic relief in the cinema, see Anthony Balducci, *The Funny Parts: A History of Film Comedy Routines and Gags* (Jefferson, NC, 2012), pp. 70–74.

2 GOOD SHOT: EUROPE AND AMERICA MEET THE GORILLA

1 Susan Sontag, *On Photography* (New York, 1977), p. 13.

2 Thomas S. Savage, 'Notice of the External Characters and Habits of *Troglodytes Gorilla*, a New Species of Orang from the Gaboon

River'; Jeffries Wyman, 'Osteology of the Same', *Boston Journal of Natural History*, v/4 (December 1847), p. 420.

3 Isidore Geoffroy Saint-Hilaire, 'Note sur le gorille', *Annales des sciences naturelles. Zoologie*, ser. 3, xvi (Paris, 1851), pp. 154–8.

4 P.-A. Cap, *Le Muséum d'histoire naturelle* (Paris, 1854), p. 228.

5 'Wild Men of the Woods: The Gorilla', *The Leisure Hour: A Family Journal of Instruction and Recreation*, 370 (27 January 1859), p. 57.

6 'The Gorilla', *The Illustrated London News* (9 April 1859), p. 348.

7 Emmanuel Frémiet, interviewed in Thiébault-Sisson, 'Au jour le jour, une vie d'artiste : Emmanuel Fremiet', *Le Temps* (3 January 1896), reprinted in Catherine Chevillot, *Emmanuel Fremiet, 1824–1910: La Main et le multiple*, exh. cat., Musée des Beaux-Arts de Dijon (1988), p. 187.

8 Gauthier Laboulaye, quoted in M. Dureau de la Malle, 'Mémoire sur le grand gorille de Gabon, *Troglodytes gorilla*, déterminant la limite de la navigation d'Hannon, le long des côtes de l'Afrique occidentale', *Annales des sciences naturelles. Zoologie*, ser. 3, xvi (Paris, 1851), p. 191.

9 'The Gorillas at the British Museum', *The Times* (26 July 1861).

10 Michel Vaucaire, *Paul du Chaillu: Gorilla Hunter*, trans. Emile Pepper Watts (New York and London, 1930), p. 135.

11 Paul B. du Chaillu, *Explorations and Adventures in Equatorial Africa* (New York, 1861), pp. 84, 86, 98, 101, 487.

12 'Scientific Fracas', *The Times* (4 July 1861); 'Latest from Africa', *Punch, or the London Charivari* (13 July 1861).

13 Paul B. du Chaillu, *A Journey to Ashango-Land: And Further Penetration into Equatorial Africa* (London, 1867), p. 2.

14 Joan M. Dixon, 'Melbourne 1865: Gorillas at the Museum', in *A Museum for the People: A History of the Museum of Victoria and its Predecessors, 1854–2000*, ed. Carolyn Rasmussen (Melbourne, 2001), pp. 68–9.

15 R. M. Ballantyne, *The Gorilla Hunters* (London and New York, 1861), p. 118.

16 Bula N'Zau (pseudonym of Henry Bailey), *Travel and Adventures in the Congo Free State and its Big Game Shooting* (London, 1894),

p. 10. A comprehensive account of the history of shooting gorillas was first provided by Geoffrey H. Bourne and Maury Cohen, *The Gentle Giants: The Gorilla Story* (New York, 1975), pp. 126–74.

17 Fritz Duquesne, 'Hunting Ahead of Roosevelt: The Ugly Rhinoceros and Smaller Game', *The Ogden Standard* (Utah, 12 June 1909); Etienne Bazin, 'Hunting Ahead of Roosevelt: Rogue Elephants and Other Rogues', *The Ogden Standard* (Utah, 27 November 1909); 'A Duel Between Brutes: Extraordinary Contest Between a Gorilla and a Lion', *The Omaha Daily Bee* (5 March 1888); Grant Allen, 'My First Gorilla', *The Princeton Union* (14 June 1900); 'Trapping a Gorilla: Story of a Vicious Struggle in the African Jungle', *Daily Arizona Silver Belt* (10 July 1909).

18 Robert von Beringe, 'Bericht des Hauptmanns von Beringe über seine Expedition nach Ruanda', *Deutsches Kolonialblatt* [1903], in George B. Schaller, *The Mountain Gorilla: Ecology and Behaviour* (Chicago, IL, 1963), p. 15.

19 Andreas von Beringe, 'On the Trail of the Man who Discovered the Mountain Gorilla', *Gorilla Journal* (24 June 2002), www.berggorilla.org.

20 Schaller, *The Mountain Gorilla*, p. 15 and Table 1.

21 Sir Harry Johnston, introduction to T. Alexander Barns, *The Wonderland of the Eastern Congo* (London and New York, 1922), p. xxv. A similar indignity was inflicted upon the bodies of giant chimpanzees, such as those illustrated strung up in Adolf Friedrich Duke of Mecklenburg's *From the Congo to the Niger and the Nile* (1913), and Sir Harry Johnston's *George Grenfell and the Congo* (1908).

22 Prince William of Sweden, *Among Pygmies and Gorillas: With the Swedish Zoological Expedition to Central Africa 1921* (London, 1923), pp. 15, 40, 195.

23 T. Alexander Barns, *Across the Great Craterland to the Congo* (London, 1923), pp. 134, 135, 145.

24 T. Alexander Barns, *An African Eldorado: The Belgian Congo* (London, 1926), pp. 55, 206.

25 Barns, writing in the *African World* (London, 1 October 1924), quoted in Mary L. Jobe Akeley, *Carl Akeley's Africa: The Account of*

the Akeley – Eastman – Pomeroy African Hall Expedition (London, 1931), p. 244.

26 William King Gregory, 'In Quest of Gorillas, I: On Our Way to Gorilla-Land', *The Scientific Monthly*, XLI/5 (November 1935), p. 387.

27 William King Gregory, 'In Quest of Gorillas, V: Elusive Giants of the Mountains', *The Scientific Monthly*, XLII/3 (March 1936), p. 274.

28 William King Gregory, 'In Quest of Gorillas, X: Cameroon Folks', *The Scientific Monthly*, XLIII/2 (August 1936), p. 139.

29 Attilio Gatti, 'Among the Pygmies and Gorillas', *Popular Mechanics*, LVIII/3 (September 1932), p. 420. Attilio Gatti, 'Gorilla!', *Boys' Life* (May 1932), pp. 16, 45–6.

30 Attilio Gatti, *Great Mother Forest* (London, 1936), pp. 106, 116.

31 George Merfield, with Harry Miller, *Gorillas Were My Neighbours* (London, 1956), p. 62. The French translation of Merfield's memoirs was, tellingly, titled *Good bye gorilles (*Paris, 1957). It has been noted that the Powell Cotton Museum at Berchington in Kent contains the remains of 217 gorillas; see A. F. Dixson, *The Natural History of the Gorilla* (London, 1981), p. 8.

32 Merfield, *Gorillas Were My Neighbours*, pp. 168, 214.

33 'If I showed that ladies with no previous hunting experience could hunt gorillas, elephants, and lions, much of the heroics which have attached to African big-game hunting would begin to wane.' Carl Akeley, *In Brightest Africa* (Garden City, NY, 1923), p. 226.

34 Mary Hastings Bradley, *On the Gorilla Trail* (New York, 1923), p. 4.

35 Akeley, *In Brightest Africa*, p. 221. Akeley's film, *Meandering in Africa*, is held today in the American Museum of Natural History Research Library.

36 Carl Akeley, 'My Hunt for the Mountain Gorilla', in Carl and Mary L. Jobe Akeley, *Lions, Gorillas and Their Neighbours* (London, 1932), p. 167.

37 Akeley, *In Brightest Africa*, pp. 114, 216.

38 Carl Akeley to the Bradleys, 11 July 1921, cited in Penelope Bodry-Sanders, *Carl Akeley: Africa's Collector, Africa's Savior* (New York, 1991), p. 178.

39 Akeley, *In Brightest Africa*, pp. 239, 242.
40 Clyde Fisher, 'Carl Akeley and His Work', *The Scientific Monthly*, xxiv/2 (February 1927), p. 109.
41 Akeley, *In Brightest Africa*, p. 265.
42 Jeannette Eileen Jones, '"Gorilla Trails in Paradise": Carl Akeley, Mary Bradley, and the American Search for the Missing Link', *The Journal of American Culture*, xxix/3 (September 2006), p. 328.
43 'Church Will Allow Akeley's "Chrysalis": Statue Suggesting Evolution, Rejected by National Academy, Invited by a Unitarian Pastor', *The Art News*, xxii/27 (12 April 1924), p. 4.
44 See W. R. Leigh, *Frontiers of Enchantment: An Artist's Adventures in Africa* (New York, 1938).
45 Henry Fairfield Osborn, Foreword to Akeley, *In Brightest Africa*, pp. xi–xii.
46 Carl Akeley, quoted in Mary L. Jobe Akeley, *Carl Akeley's Africa*, p. 243.
47 Frank Parker Stockbridge, '"It's Brains Against Brains – Fighting Gorillas!" A Famous Big Game Hunter Tells of Tricks He Devised to Capture the "Beast That Fights Like a Man"', *Popular Science Monthly* (December 1926), p. 13. See also Ben Burbridge, *Gorilla: Tracking and Capturing the Ape-man of Africa* (London, 1928).
48 Historian Georgina Montgomery has succinctly captured his conflicting motivations in this regard: 'At times Yerkes emphasized his emotional attachment to Congo, a primate he was exceptionally excited to study. As his research unfolded, however, Yerkes became disappointed by Congo's apparent lack of intelligence and distanced her from humans.' Georgina M. Montgomery, '"Infinite Loneliness": The Life and Times of Miss Congo', *Endeavour*, xxxi–i/3 (July 2009), p. 105.
49 This *New York Times* review provides an excellent summary of this hard-to-locate film's gorilla content: Mordaunt Hall, 'African Gorillas and Savages', *New York Times* (14 December 1926).
50 Ben Burbridge, 'Kidnapping Giant Gorillas in the Congo', *The Milwaukee Journal* (6 January 1929).
51 'Congo's End', *Time* (7 May 1928).

52 Martin Johnson, *Congorilla: Adventures with Pygmies and Gorillas in Africa* (New York, 1931), pp. 160, 132–3, 246, 278, 281.

53 Schaller, *The Mountain Gorilla*.

54 Ibid., p. 2.

55 Ibid., p. 23.

56 Harold C. Bingham, *Gorillas in a Native Habitat; Report of the Joint Expedition of 1929–30 of Yale University and Carnegie Institution of Washington for Psychobiological Study of Mountain Gorillas (Gorilla beringei) in Parc national Albert, Belgian Congo, Africa* (Washington, DC, 1932).

57 'Thinks Well of Gorillas. Prof. Richard L. Garner Tells of his Life in a Cage', *New York Times* (26 March 1894); Richard L. Garner, *Gorillas and Chimpanzees* (London, 1896).

58 Schaller, *The Mountain Gorilla*, pp. 290, 292.

59 George B. Schaller, *The Year of the Gorilla* [1964] (Chicago, IL, 2010), pp. 34, 35.

60 Ibid., pp. 91, 211.

61 'Record Shipment of Young Apes Swells U.S. Gorilla Population', *Life*, XI/12 (22 September 1941), pp. 66–7.

62 '*Life* Goes on a Gorilla Hunt: A Big "Garçon" is Killed Protecting His Family. Zoos Keep Said Working', *Life*, XXXI/21 (19 November 1951), pp. 175–80.

63 'Gorilla Hunt (Letter to the Editor)', *Life*, XXXI/24 (10 December 1951), p. 16.

64 Schaller, *The Year of the Gorilla*, p. 201.

65 Ibid., p. 211.

66 Dian Fossey, 'Making Friends with Mountain Gorillas', *National Geographic*, CXXXVII/1 (January 1970), p. 51.

67 Ibid., p. 65.

68 Ibid.

69 Dian Fossey, *Gorillas in the Mist* (Harmondsworth, 1983), p. 105.

70 Schaller, *The Mountain Gorilla*, p. 201; Fossey, *Gorillas in the Mist*, p. 46.

71 Schaller, *The Mountain Gorilla*, p. 330.

72 Fossey, *Gorillas in the Mist*, p. 183.

73 Ibid., p. 103.
74 Ibid., p. 209.
75 Camilla de la Bédoyère, *No One Loved Gorillas More: Dian Fossey's Letters from the Mist* (Washington, DC, 2005), p. 160.
76 Dian Fossey, 'The Imperiled Mountain Gorilla: A Grim Struggle for Survival', *National Geographic*, CLIX/4 (April 1981), p. 501.
77 Fossey, *Gorillas in the Mist*, p. 57.
78 A. H. Harcourt and K. J. Stewart, 'Functions of Wild Gorilla "Close" Calls: Repertoire, Context and Interspecific Comparison', *Behaviour*, CXXXIII (1996), pp. 827–46; cited in Andrew Y. Grant, *Nearly Human: The Gorilla's Guide to Good Living* (Suffern, NY, 2007), pp. 82–3.
79 Richard W. Byrne, 'Clever Hands: The Food-processing Skills of Mountain Gorillas', in *Mountain Gorillas: Three Decades of Research at Karisoke*, ed. Martha M. Robbins, Pascale Sicotte and Kelly J. Stewart (Cambridge, 2001), pp. 293–313.
80 Revenue from the permits for foreign tourists is channelled into conservation programmes and the upkeep of the park, as well as a small amount being devoted to projects for local people. See 'Gorilla Trekking', www.rwandatourism.com, accessed 27 March 2011.
81 See, for example, Stephanie McCrummen, 'For Congo's Gorillas the War Goes On', *The Guardian Weekly* (3 August 2007), p. 28; Scott Johnson, 'Gorilla Warfare', *Newsweek* (6 August 2007), pp. 25–9; Mark Jenkins, 'Who Murdered the Virunga Gorillas?', *National Geographic*, CCXIV/1 (July 2008), pp. 34–65; Paul Raffaele, 'Guerillas in Their Midst', *The Weekend Australian* (*Good Weekend* magazine, 11 October 2008), pp. 51–7.

3 GORILLA MANIA

1 Filippo De Filippi, *L'Uomo e le scimie* (Man and the Apes, Turin, 2009).
2 *Punch* magazine amused its readers with an 'alarmed flunkey', nervously announcing the arrival of 'MR G-G-G-O-O-O-RILLA!' at an elegant London soirée. The music sheet illustrators Concanen and

Lee satirized him as a dinner-suited gorilla performer conducting C.H.R. Marriott's catchy new 'Gorilla Quadrille'. Another of their fantasies, 'Mr Gorilla: The Lion of the Season', showed a suave primate belting out the lyrics to a song by the popular entertainer Howard Paul: 'Ladies and Gentlemen, how do you do / I am the Gorilla of Monsieur Chaillu'.

3 Michel Vaucaire, *Paul du Chaillu: Gorilla Hunter* (New York and London, 1930), p. 217.

4 For a full account of these and other jungle heroines, see Don Markstein's comprehensive Toonopedia website at www. toonopedia. com. See also the Gorilla Cover Gallery at the Comic Book Gorillarama website at members.shaw.ca. Both accessed 12 March 2011.

5 See the Grand Comics Database at www.comics.org; and the DC Comics Database at www.dc.wikia.com. Both accessed 14 March 2011.

6 'Konga Meets the Creatures from Beyond Space', *Konga*, 1/7 (July 1962), p. 1.

7 See, for example, Andre Gaza's 'Strangled by a Giant Gorilla', *Man's Life*, VI/2 (March 1958), pp. 32–3, 60–62. For this genre, see the Classic Men's Magazine Covers site at www.stagmags.com; and the extraordinary Galactic Central site at www.philsp.com. Both accessed 8 March 2011.

8 See the articles 'Girl-crazy Gorillas, Men's Adventure Magazine Style' and 'Looking for Artist Will Hulsey, creator of "Weasels Ripped My Flesh" and much more', posted by SubtropicBob at www.menspulpmagazines.com. Both accessed 15 February 2011.

9 On Captain Congo, see 'Captain Congo Preview', www. panicproductions.com.au, accessed 10 February 2012. On Aneurin Wright, see Adrienne Rappaport, 'Aneurin Nye Wright: Things To Do in a Retirement Home Trailer Park . . . When You Are 29 and Unemployed' in the webzine *Sequential Tarts*, at www.sequentialtarts.com, accessed 5 January 2011.

10 'Biggest Gorilla Ever Seen By Man', *Los Angeles Times* (17 August 1901).

11 Robert Hartmann, *Anthropoid Apes* (London, 1885), p. 261.
12 'A Gorilla At The Zoological Gardens', *The Times* (25 October 1887).
13 A. D. Bartlett, *Land and Water* (22 October 1887), quoted in A. D. Bartlett, *Wild Animals In Captivity*, ed. Edward Bartlett (London, 1898), p. 146.
14 'The Museum of the College of Surgeons', *The British Medical Journal*, II/1436 (7 July 1888), p. 26.
15 However, an infant female gorilla had apparently made its way into Wombwell's Travelling Menagerie as early as 1855, when it had been misidentified as a chimpanzee. See Thomas J. Moore, 'A Gorilla', *The Times* (24 June 1876).
16 'The Berlin Gorilla', *The Times* (23 July 1877).
17 'Death of "Pongo"', *The Times* (26 November 1877); 'Pongo', *The Times* (7 December 1877). Oscar Lenz visited Africa at the behest of the Berlin Geographical Society in 1872, as did Dr Paul Güssfeldt in 1873. As Professor of Geography and Geology at the University of Czernowitz, Lenz was also to lead an expedition to the Congo for the Geographical Society of Vienna in 1885.
18 See 'Gorilla and Chimpanzee At The Crystal Palace', *The Graphic* (13 September 1879), p. 269.
19 'The Young Gorilla at the Jardin des Plantes', *New York Times* (29 June 1884).
20 'A Gorilla in Town', *New York Times* (26 September 1867); 'Chiko to Have a Gorilla Bride', *New York Times* (16 December 1893); James A. Bailey, 'Mr Bailey Disagrees', *The World* (New York, 31 December 1893).
21 'Big Reception to a Gorilla', *The Boston Sunday Globe* (15 April 1894); 'Chiko Stuffed For Good', *The World* (New York, 16 January 1895).
22 'Johanna Dresses and Acts Just Like a Woman', *The World* (11 April 1897).
23 'Barnum's in London', *New York Times* (16 January 1898).
24 Eli Harvey, 'Little Miss Dinah of Africa', *The Art World*, II/4 (July 1917), p. 341.

25 William T. Hornaday, 'Gorillas, Past and Present', *Zoological Society Bulletin*, XVIII/1 (January 1915), p. 1185.
26 Raymond L. Ditmars, *Confessions of a Scientist* (New York, 1934), p. 226.
27 Hornaday, 'Gorillas, Past and Present', p. 1183.
28 Sylvia Loomis, Oral History Interview with Eugenie Shonnard, 27 February–9 April 1964, Archives of American Art, Smithsonian Institution. Eli Harvey, 'Little Miss Dinah of Africa', pp. 341–2.
29 Anon., 'The Zoological Park Gets a Gorilla', *Zoological Society Bulletin*, XVII/5 (September 1914), p. 1130.
30 Richard L. Garner, 'Gorillas in Their Own Jungle', *Zoological Society Bulletin*, XVII/3 (May 1914), p. 1104.
31 Hornaday, 'Gorillas, Past and Present', p. 1183.
32 'Jungle Baby Lolls in Invalid's Luxury. Bronx Park Gorilla is Trundled in Perambulator by a 225-Pound Keeper', *New York Times* (21 December 1914).
33 'A Tame Gorilla', *The Times* (7 August 1920).
34 Alyse Cunningham, 'A Gorilla's Life in Civilization', *Zoological Society Bulletin*, XXIV/3 (September 1921), p. 123.
35 'Gorilla Brings Secretary: Circus Attraction Due on Liner Today Traveling de Luxe', *New York Times* (22 March 1921).
36 'Grieving Gorilla Dead at Garden', *New York Times* (18 April 1921).
37 'Girl Crossing Sea to Tend Circus Ape', *New York Times* (17 April 1921); 'John Daniels [sic] to have Nurse; Appeal to London for Gorilla', *New York Tribune* (17 April 1921); 'A Human Gorilla: Killed By Change of Air', *The Times* (19 April 1921); 'Specialists Study John Daniel's Body', *New York Times* (25 April 1921).
38 'Another Gorilla is on His Way Here: "Cousin" of John Daniel, Accompanied by Woman Guardian, to Join Circus Saturday. "He's a John Daniel Too". Miss Alyce [sic] Cunningham, Who Trained His Predecessor, to Be His Constant Companion', *New York Times* (3 April 1924); 'Akeley Will Greet Gorilla Down Bay: Discoverer of Largest Apes' Mildness in Wilds Eager to Meet John Daniel 2nd', *New York Times* (4 April 1924); 'J. Sultan, Gorilla, Hotel Guest Here: Successor to John Daniel, With

Corrected Name, Praised by Naturalist for Intelligence', *New York Times* (7 April 1924); 'John, the Gorilla, Bites His Mistress', *New York Times* (8 April 1924).

39 Gwynne Vevers, *London's Zoo* (London, 1976), p. 96.

40 'Trained Gorilla For the Zoo: Miss Cunningham's Ape', *The Times* (9 May 1925).

41 Frank Buck, *Bring 'Em back Alive* (London, 1931), p. 146.

42 Hornaday, 'Gorillas, Past and Present', p. 1185.

43 Ellen Velvin, *From Jungle to Zoo* (London, 1914), p. 90.

44 Alexander Sokolowsky, *Beobachtungen über die Psyche der Menschenaffen* ('Observations on the Psyche of the Great Apes', Frankfurt, 1908), cited in Nigel Rothfels, *Savages and Beasts: The Birth of the Modern Zoo* (Baltimore, MD, 2007), p. 3.

45 Raymond L. Ditmars and William Bridges, *Wild Animal World: Behind the Scenes at the Zoo* (New York, 1937), p. 146.

46 'Mok and Moina of the Zoo: Young Gorillas' First Winter', *The Times* (18 March 1933); 'Mok and Moina Move In: Delight At New Abode', *The Times* (29 April 1933); 'Death of Mok: Gorilla's Mysterious Illness', *The Times* (15 January 1938).

47 'Baby Gorilla at London Zoo: An Ear for French', *The Times* (7 November 1947); 'Boy Bitten By Gorilla: London Zoo Incident', *The Times* (19 June 1956); 'Rare Gorilla for London Zoo: Deserted by Mother in Uganda Forest', *The Times* (19 May 1960); 'Zoo Gorilla Bites Visitor: Official Says Man Climbed Barrier', *The Times* (12 September 1960).

48 J. Barrington-Johnson, *The Zoo: The Story of London Zoo* (London, 2005), p. 142.

49 A. F. Dixson, H.D.M. Moore and W. V. Holt, 'Testicular Atrophy in Captive Gorillas', *Journal of Zoology*, CXCI/3 (July 1980), pp. 315–22. Tony Samstag, 'To Guy With Gratitude', *The Times* (5 November 1982).

50 'Hello Susie', *The Ironwood Times* (Michigan, 24 May 1929); 'Zeppelin is Forced Back on U.S. Trip', *The Lethbridge Herald* (16 May 1929); 'Disabled Motors Force Zeppelin Back', *The Olean Herald* (New York, 16 May 1929.

51 'German Zeppelin May Not Reach Lakehurst Till Monday Morning', *The Lethbridge Herald* (3 August 1929).
52 'Zeppelin's Atlantic Trip is Successful, Landing on Sunday', *The Lethbridge Herald* (5 August 1929).
53 'Bushman the Gorilla has Birthday in Chicago Zoo', *Life* (9 June 1941), pp. 97–8.
54 'Manners and Morals: The Jovial Gorilla', *Time* (26 June 1950).
55 Gene Plowden, *Gargantua: Circus Star of the Century* (New York, 1972), p. 27.
56 Fred Bradna, *The Big Top: My Forty Years with the Greatest Show On Earth* (London, 1953), p. 120; 'Jungle to Garden', *Time* (18 April 1938).
57 'From Day To Day', *Cairns Post* (8 April 1939).
58 'Bold Photographer Outsmarts "Gargantua the Great"', *Life*, IV/11 (14 March 1938), pp. 20–21; J.B.T. Scripps, 'Gargantua: World's Most Successful Animal Lives For One Purpose: Murder', *Life*, VIII/9 (February 1940), pp. 63–4, 80–83.
59 Robert Lewis Taylor, *Center Ring: The People of the Circus* (Garden City, NY, 1956), pp. 198, 201.
60 'A Wedding Has Been Arranged', *The Mail* (Adelaide, 8 August 1942).
61 Gene Plowden, *Gargantua: Circus Star of the Century* (New York, 1972), p. 85; 'Gargantua, Mighty Circus Gorilla, Dies in His Cage', *Freeport Journal-Standard* (Illinois, 26 November 1949), front page; 'Gargantua Dies in Miami with a Snarl On His Face', *The Miami Herald* (26 November 1949), front page.
62 Alan F. Dixson, *The Natural History of the Gorilla* (London, 1981), p. 9.
63 L. M. Boston, *A Stranger at Green Knowe* (London, 1961).
64 Mark Prigg, 'London Zoo Mourns Bobby the Poster Boy', *Evening Standard* (5 December 2008).
65 See Michael McRae, 'Central Africa's Orphan Gorillas: Will They Survive in the Wild?', *National Geographic* (February 2000), pp. 84–97.
66 *Kwibi and Damian Aspinall Gorilla School*, Aspinall Foundation's posting on YouTube at www.youtube.com; Aqua Vita Films

promo, viewed over 1.87 million times: *Gorilla Reunion: Damian Aspinall's Extraordinary Gorilla Encounter on Gorilla School*, at www.youtube.com, both accessed 29 April 2011.

4 SEX AND CRIME

1 Norman Macleod, 'The Gorilla', in *Good Words for 1861*, ed. Norman Macleod (London, 1861), p. 468.
2 See Marek Zgórniak, 'Frémiet's *Gorillas*: Why Do They Carry Off Women?', *Artibus et Historiae*, XXVII/54 (2006), pp. 219–37.
3 All quotes from 'Comstock Objects: He Says the Posters of the Gorilla and the Girl Violate the Law', *The World* (New York, 28 November 1893).
4 'Eden Musee', *The Evening World* (New York, 5 December 1893).
5 '"The Gorilla" Shows at Orpheum Theater Today', *Titusville Herald* (Pennsylvania, 20 December 1930).
6 'Mystery Comedy an Instant Hit at the Academy', *The Daily Mail* (Hagerstown, Maryland, 30 December 1927).
7 'Gorilla Man Brands L.A. Beauty', *Los Angeles Evening Herald* (9 June 1927); 'Brands Young Actress With Razor Blade: Gorilla-like Fiend Prints Scarlet "K" Seven Times', *The Bakersfield Californian* (9 June 1927); 'Tale of Gorilla Man's Attack on Hollywood Actress Investigated', *Los Angeles Times* (10 June 1927).
8 'Indiana Town Believes Has Gorilla Man', *The Bakersfield Californian* (17 January 1927); 'Suspected Strangler Held in East', *Oakland Tribune* (17 January 1927).
9 'How They Tracked His Murders Home to "Gorilla Man": Modern Science and Extraordinary Detective Work Trap at Last the Strangler Who Spread Terror Over a Dozen States and Fasten to Him the Deaths of Twenty-One Women and Girls Who Gasped Out Their Lives Under the Grip of His Ape-like Hands', *San Antonio Light* (11 December 1927).
10 '$1750 Is Awarded for Embraces of "Gorilla Man"', *The Bakersfield Californian* (30 January 1932); 'Hugged By Gorilla Man, Woman Asks $302,300', *Chester Times* (Pennsylvania, 13 December 1929).

Having sued for $302,300, Mrs Pasley was eventually awarded $1,750 for her gorilla 'trauma'.

11 'Execution of "Gorilla Man" Denied', *El Paso Herald* (Texas, 11 October 1956); '"Gorilla Man" Killer Is Reported Insane', *Oakland Tribune* (California, 12 March 1932).

12 Robert Graysmith, *The Laughing Gorilla: The True Story of the Hunt for One of America's First Serial Killers* (New York, 2009), pp. 68, 332–9.

13 'Gorilla-man Escapes', *Woodland Democrat* (California, 16 January 1934); 'Alleged "Gorilla Man" Confesses to Murder of Woman', *Corsican Daily Sun* (Texas, 22 September 1936); 'Murder Confessed By Gorilla Man: Donald Hazell Tells of Guilt in Ruth Muir Case. 225-Pound Giant Claims Psychic Spell Forced Him to Slay', *The Bakersfield Californian* (22 September 1936); 'Gorilla Man in New Attack', *San Mateo Times* (California, 23 May 1938); '"Gorilla Man" in Chicago Attacks, Kills New Victim', *The Bakersfield Californian* (27 May 1938).

14 '"Wild Man" May Be a Gorilla', *New York Times* (30 August 1895); 'There's No Gorilla Near White Plains', *New York Times* (22 November 1894).

15 'The Gorilla Objected', *New York Times* (18 November 1888); 'Gorilla Out Of Its Cage: Barnum & Bailey's Animal Creates a Scare at Binghamton', *New York Times* (16 May 1893).

16 'S.W. Iowa Farmers Still Fear Gorilla', *The Oxford Leader* (Iowa, 20 November 1930).

17 Alexander Woolcott, 'The Hairy Ape, a play in eight scenes, by Eugene O'Neill', *New York Times* (10 March 1922).

18 A good summary of George Barrows's career as a gorilla actor is given by Bob Burns in *Hollywood Goes Ape!* (1994), recently made available on the compilation DVD *Dinosaurs vs. Apes*, released by Frontline Entertainment and Cinema Epoch in 2007.

19 'Chicago Zoo Gorilla Gets Loose; Garter Snake Scares Him Back', *Wisconsin State Journal* (2 October 1950).

20 'Huge Gorilla Is Recaptured at Hot Springs', *Northwest Arkansas Times* (Fayetteville, 24 April 1951).

21 The English singer-songwriter Jake Thackray translated Brassens' classic as 'Brother Gorilla' in 1972.
22 'Gorilla On The Lam Terrifies a Town', *Garden City Telegram* (Kansas, 16 May 1956); 'Gorilla Escapes', *Corsicana Daily Sun* (17 May 1956); 'Gorilla Bites Tampa Man', *The Lowell Sun* (8 November 1956); 'Deputy Kills Gorilla With Single Shot', *Fort Pierce News-Tribune* (Florida, 8 November 1956).
23 'Escaped Gorilla Frightens Antwerp; Finally Shot', *Ames Daily Tribune* (Iowa, 19 July 1958); 'Seven Foot Gorilla Killed on Housetop', *The Tipton Daily Tribune* (Indiana, 19 July 1958).
24 A summary of these and other gorilla escapes is provided by Sherry Jacobson, 'Dallas Zoo Says Response to Gorilla Escape Was By the Book', *Dallas Morning News* (7 March 2010). See also 'Franklin Park Gorilla Escapes, Attacks 2', *The Boston Globe* (29 September 2003); and 'Texas: Gorilla Escapes and Injures 4 at Zoo', *New York Times* (20 March 2004).
25 'Four Hurt as Gorilla Escapes at Dutch Zoo', *The Guardian* (London, 18 May 2007); 'King Kong, Gone Wrong and On the Run, Terrorises Zoo Visitors', *Sydney Morning Herald* (20 May 2007).
26 'The Gorilla Film "Ingagi"', *Science*, LXXI/1849 (1930), p. x.
27 T.B.R., 'The Passing World', *Manitoba Free Press* (Winnipeg, 27 December 1929).
28 One of these designs, dated December 1928, survives in the Musée National des Arts et Traditions Populaires, Paris (inv. 67.116.61 D).
29 Mark Cotta Vaz, *Living Dangerously: The Adventures of Merian C. Cooper, Creator of King Kong* (New York, 2005), p. 14. Cooper 'could to his last days quote at length' from Du Chaillu's writings; see George E. Turner with Orville Goldner, revised and expanded by Michael H. Price with Douglas Turner, *Spawn of Skull Island* (Baltimore, MD, 2002), p. 24.
30 Programme reproduced in Ronald Gottesman and Harry Geduld, eds, *The Girl in the Hairy Paw: King Kong as Myth, Movie and Monster* (New York, 1976), p. 8.
31 See Snead's essay 'Spectatorship and Capture in *King Kong*: The Guilty Look', in James Snead, *White Screen, Black Images:*

Hollywood from the Dark Side (New York and London, 1994), p. 22.

32 Fatimah Tobing Rony, *The Third Eye: Race, Cinema and Ethnographic Spectacle* (Durham and London, 1996), pp. 159, 180. The literature on *King Kong* is vast; see, among others, Orville Goldner and George E. Turner, *The Making of King Kong: The Story Behind a Film Classic* (London, 1975); Karen Haber, ed., *Kong Unbound: The Cultural Impact, Pop Myths, and Scientific Plausibility of a Cinematic Legend* (New York, 2005); Ray Morton, *King Kong: The History of a Movie Icon. From Fay Wray to Peter Jackson* (New York, 2005); and Paul A. Woods, ed., *King Kong Cometh!* (London, 2005).

33 Quoted in Rudy Behlmer, 'Foreword', in *The Girl in the Hairy Paw: Kong Kong as Myth, Movie and Monster*, ed. Ronald Gottesman and Harry Geduld (New York, 1976), p. 10.

34 Barbara Creed, *Darwin's Screens: Evolutionary Aesthetics, Time and Sexual Display in the Cinema* (Melbourne, 2009), pp. 190–91.

35 Ray Gill, 'High, Wired and Handsome, Kong Rises in Melbourne', *The Age* (19 September 2009), p. 5. See also www.creaturetechnology.com, accessed 10 February 2012.

36 Bernice Kanner, 'The Soft Side of Luggage', *New York Magazine* (31 October 1983), p. 22.

37 According to West enthusiast and collector Ramfiz Diaz. See Mark J. Price, 'Gorilla Jones Story a Virtual Knockout – Mae West Fan Pleased To Learn About Akron Boxer's Role in Life of Actress', *Beacon Journal* (Akron, Ohio, 29 June 2009). See also Mark J. Price, 'Akron's King of Rings – Boxer Gorilla Jones Conquers the World', *Beacon Journal* (Akron, Ohio, 8 June 2009).

38 Henry McLemore, 'Today's Sports Parade', *The Sandusky Star-Journal* (9 September 1937).

39 'Dan Cupid Gets Decision Over Burly "Bull" Montana', *The Lethbridge Herald* (20 September 1929).

40 'Gorilla-faced Earl', *Time* (7 December 1936).

41 See Bruce Moore, 'Racing Slang in Australia', www.anu.edu.au, accessed 5 May 2011.

42 'Uncle Sam's Post Office Policeman Human Gorilla With 89-inch Reach', *The Day Book* (Chicago, 12 March 1914).

43 Paul Maccabee, *John Dillinger Slept Here: A Crooks' Tour of Crime and Corruption in St Paul, 1920–1936* (St Paul, MN, 1995), p. 51.
44 Judith Shatnoff, 'A Gorilla to Remember', *Film Quarterly*, XXII/1 (Autumn 1968), p. 58.
45 'Long Voyage Home: Having Won Its War in the Pacific, the Navy Returns to have Its Day', *Life*, XIX/18 (29 October 1945), p. 40. A feature on Addams in 1953 noted that he loved gorillas, watching them for hours at the zoo. 'Charles Addams: Master of the Macabre', *Look* (19 May 1953).
46 Robert Brustein, review of *Gorilla Queen* in *The New Republic* (6 May 1967), quoted in David Kerry Heefner, 'The Gorilla Was Gay: Remembering Ronald Tavel's "Gorilla Queen", March 1967', www.outhistory.org, accessed 1 May 2011.
47 Cynthia Erb, *Tracking King Kong: A Hollywood Icon in World Culture*, 2nd edn (Detroit, MI, 2009), p. 173.

5 ALL IN THE FAMILY

1 Stephen Jay Gould, *Ever Since Darwin: Reflections in Natural History* (Harmondsworth, 1977), pp. 260–67.
2 Thomas Henry Huxley, *Evidence as to Man's Place in Nature* (London, 1863). For these and later quotes, see pp. 57, 60, 70, 71, 84, 104, 105–6.
3 Owen wanted the museum to be an encyclopaedic monument to 'natural theology' that would awe visitors by showing the fullest extent of God's majesty in creation, while Huxley argued for only a small public display of appealing specimens, with the majority of the collections kept in storage for serious study by the scientific community. See Carla Yanni, 'Divine Display or Secular Science: Defining Nature at the Natural History Museum in London', *The Journal of the Society of Architectural Historians*, LV/3 (September 1996), pp. 276–99.
4 Richard Owen, *Memoir on the Gorilla (Troglodytes Gorilla, Savage)* (London, 1865), p. 52.
5 'Monkeyana', *Punch; or, the London Charivari* (18 May 1861), p. 206.

6 Frederick McCoy, quoted in 'The Gorillas at the Museum', *The Australian News for Home Readers* (25 July 1865), p. 7.
7 'Artistic Jottings at the National Museum', *The Australasian Sketcher* (5 August 1876), p. 67. On the 'Gorilla Wars' in Melbourne, which are representative of the evolutionary skirmishes that broke out in major cities worldwide in the mid-1860s, see Barry W. Butcher, 'Gorilla Warfare in Melbourne: Halford, Huxley and "Man's Place in Nature"', in *Australian Science in the Making*, ed. R. W. Home (Cambridge, 1988), pp. 153–69.
8 Huxley, *Evidence as to Man's Place in Nature*, p. 109.
9 Janet Browne, 'Charles Darwin as a Celebrity', *Science in Context*, XVI/1–2 (Spring–Summer 2003), p. 186.
10 See L. Perry Curtis Jr, *Apes and Angels: The Irishman in Victorian Caricature* (Washington, DC, 1971).
11 Will Dyson, *Kultur Cartoons*, with a foreword by H. G. Wells (London, 1915). Ross McMullin, *Will Dyson: Cartoonist, Etcher and Australia's Finest War Artist* (London, Sydney, Melbourne, 1984), pp. 113–17.
12 Peter Fullerton, *Norman Lindsay: War Cartoons, 1914–1918* (Melbourne, 1983), p. 4.
13 We are grateful to Peter Zegers, Research Fellow at the Art Institute of Chicago, for informing us about these works. For more images of the gorilla in Second World War propaganda, see Zeger's forthcoming publication *Windows on the War: Soviet TASS Posters at Home and Abroad, 1941–1945* (Art Institute of Chicago). The full run of *Kladderadatsch* has been digitized at www.ub.uni-heidelberg.de, accessed 12 February 2011.
14 Edgar Rice Burroughs, *Tarzan of the Apes* [original magazine publication 1912, first publication in book form 1914] (London, 2008), pp. 39, 62, 65.
15 Jeffery E. Nash and Anne Sutherland, 'The Moral Elevation of Animals: The Case of "Gorillas in the Mist"', *International Journal of Politics, Culture, and Society*, V/1 (Autumn 1991), pp. 114, 120.
16 See discussion between film-maker John Landis, make-up artist Rick Baker and cinema gorilla historian Bob Burns, part of the

Special Edition Features on The Criterion Collection's 2011 DVD reissue of *Island of Lost Souls*, Criterion Collection no. 586.

17 'Expects to Replace All Vital Organs: Dr Voronoff Announces Startling Discovery in Substitutions from Chimpanzees', *New York Times* (20 June 1922). This article reprinted information drawn from the *Chicago Tribune*.

18 Serge Voronoff, *The Conquest of Life*, trans. G. Gibier Rambaud (London, 1928), p. 143.

19 'Brief Notices: *Sex Glands Function and the Human Life* by C. Leventis', *The Quarterly Review of Biology*, VI/3 (September 1931), p. 372.

20 Il'ya Ivanov, 1925, cited in Kirill Rossiianov, 'Beyond Species: Il'ya Ivanov and His Experiments on Cross-breeding Humans with Anthropoid Apes', *Science in Context*, XV/2 (2002), p. 289.

21 David Hamilton, *The Monkey Gland Affair* (London, 1986), pp. 67–8. The full text of Gayton's *The Gland Stealers* is available at www.archive.org, accessed 16 April 2011.

22 See 'The Gorilla Girl', *New York Magazine*, XXVII/45 (14 November 1994), pp. 36–8.

23 *Strange Adventures*, 45, online at http://dc.wikia.com, accessed 20 October 2012.

24 George B. Schaller, *The Mountain Gorilla: Ecology and Behaviour* (Chicago, IL, 1963), p. 200.

25 George B. Schaller, *The Year of the Gorilla* [1964] (Chicago, 2010), p. 181.

26 See Harry Beatty, *Journal of Mammalogy* [1951], cited in Frans de Waal, *The Ape and the Sushi Master: Cultural Reflections of a Primatologist* (New York, 2001). See also Benjamin B. Beck, *Animal Tool Behavior: The Use and Manufacture of Tools by Animals* (New York, 1980).

27 See Thomas Breuer, Mireille Ndoundou-Hockemba and Vicki Fishlock, 'First Observation of Tool Use in Wild Gorillas', *PLOS (Public Library of Science) Biology*, III/11 (November 2005), pp. 2041–3; and Liza Gross, 'Wild Gorillas Handy with a Stick',

PLOS *Biology*, III/11 (November 2005), pp. 1841–2. Both at www.plosbiology.org.

28 Gillian Sebestyen-Forrester, quoted in John Pickrell, 'Wild Gorillas Reveal Their Use of Tools', *New Scientist* (30 September 2005), www.newscientist.com.

29 See Katharine Sanderson, 'Gorillas in the Missionary Position', *Nature*, published online 14 February 2008, www.nature.com.

30 Richard Bath, 'Gana the Gorilla: Grieving Mother?', *Scotland on Sunday* (24 August 2008), at http://scotlandonsunday.scotsman.com, accessed 1 December 2010.

31 For news coverage of Binti Jua's rescue, see 'Binta Jua: Gorilla Heroine' (30 December 2005), www.bbc.co.uk; 'Gorilla Rescues Child: The World Goes Ape', part of Year in Review: 1996, http://edition.cnn.com. Amateur video of the two incidents can be viewed online: 'Jambo the Gentle Giant Jersey Zoo Boy Fell into Gorilla Pit', at www.youtube.com, accessed 30 March 2012. For a recent blog entry on gorilla emotions on the site of the Virunga National Park, see LuAnne, 'Mourning the Dead: Do Gorillas Feel Emotion?' (18–24 April 2011), http://gorillacd.org.

32 See Gregory Radick, 'Primate Language and the Playback Experiment, in 1890 and 1980', *Journal of the History of Biology*, XXXVIII/3 (Autumn 2005), pp. 461–93.

33 'The "Monkey Man": Professor Garner Returns From His Trip to Africa', *The Record-Union* (Sacramento, 26 March 1894). Prior to his expedition, Garner had written excitedly of how 'A unique and marvellous experiment among the many which I expect to be able to perform, is that of phonographing the sounds of the apes at a distance from my cage, where my phonograph will be located at times'. R. L. Garner. 'What I Expect To Do in Africa', *The North American Review*, CLIV/427 (June 1892), p. 714.

34 'Garner: The Ape Sharp. Back from a Trip to the Land of Gorillas with Lots of Simian News', *The World* (New York, 23 March 1894).

35 Emilie Genty, Thomas Breuer, Catherine Hobaiter and Richard W. Byrne, 'Gestural Communication of the Gorilla (*Gorilla gorilla*): Repertoire, Intentionality and Possible Origins', *Animal Cognition*,

XII (2009), pp. 527–46; Simone Pika, 'The Gestural Communication of Gorillas' in *The Gestural Communication of Monkeys and Apes*, ed. M. Tomasello and J. Call (New York, 2007).

36 See for example Amy S. Pollick and Frans B. M. de Waal, 'Ape Gestures and Language Evolution', *Proceedings of the National Academy of Sciences of the United States of America*, CIV/19 (8 May 2007), pp. 8184–9, www.pnas.org.

37 Presentation at American Society of Primatologists Annual Conference 2009, see www.koko.org, accessed 1 May 2011.

38 F. G. Patterson and R. Cohn, 'Self-recognition and Self-awareness in Lowland Gorillas', in *Self-awareness in Animals and Humans: Developmental Perspectives*, ed. S. T. Parker and R. W. Mitchell (New York, 1994), pp. 273–90.

39 See Gary Indiana, 'Barbet and Koko: An Equivocal Love Affair', www.criterion.com, accessed 4 April 2011.

40 On this issue, see Francine Patterson and Wendy Gordon, 'The Case for the Personhood of Gorillas', in *The Great Ape Project*, ed. Paola Cavalieri and Peter Singer (New York, 1993), pp. 58–77.

41 Francine Patterson, 'Conversations With a Gorilla', *National Geographic*, CLIV/4 (October 1978), p. 465.

42 Lee Glendinning, 'Spanish Parliament Approves "Human Rights" for Apes', *Guardian* (Thursday 26 June 2008), www.guardian.co.uk.

43 See notably Donna Haraway, *When Species Meet* (Minneapolis, MN, 2008); Jacques Derrida, *The Animal That Therefore I Am* [2006], trans. David Wills (New York, 2008); and Giorgio Agamben, *The Open: Man and Animal* [2002], trans. Kevin Attell (Palo Alto, CA, 2004).

Select Bibliography

Akeley, Carl, *In Brightest Africa* (Garden City, NY, 1923)

Bourne, Geoffrey H., and Maury Cohen, *The Gentle Giants: The Gorilla Story* (New York, 1975)

Caldecott, Julian, and Lera Miles, *World Atlas of Great Apes and their Conservation* (Berkeley, CA, 2005)

Chaillu, Paul Belloni du, *Explorations and Adventures in Equatorial Africa* (New York, 1861)

Eckart, Gene, and Annette Lanjouw, *Mountain Gorillas: Biology, Conservation, and Coexistence* (Baltimore, MD, 2008)

Erb, Cynthia, *Tracking King Kong: A Hollywood Icon in World Culture*, 2nd edn (Detroit, MI, 2009)

Fossey, Dian, 'Making Friends with Mountain Gorillas', *National Geographic*, CXXXVII/1 (January 1970)

—, *Gorillas in the Mist* (Harmondsworth, 1983)

Garner, Richard Lynch, *Gorillas and Chimpanzees* (London, 1896), available at www.archive.org

Giles-Vernick, Tamara, and Stephanie Karin Rupp, 'Visions of Apes, Reflections on Change: Telling Tales of Great Apes in Equatorial Africa', *African Studies Review*, XLIX/1 (April 2006)

Goldner, Orville, and George E. Turner, *The Making of King Kong: The Story Behind a Film Classic* (London, 1975)

Gott, Ted, '"It is lovely to be a gorilla, sometimes": The Art and Influence of Emmanuel Frémiet, Gorilla Sculptor', in *Art, Site and Spectacle: Studies in Early Modern Visual Culture*, ed. David R. Marshall (Melbourne, 2007)

—, and Kathryn Weir, *Kiss of the Beast: From Paris Salon to King Kong*, exh. cat., Queensland Art Gallery, Brisbane (2005)
Gottesman, Ronald, and Harry Geduld, eds, *The Girl in the Hairy Paw: King Kong as Myth, Movie and Monster* (New York, 1976)
Grant, Andrew Y., *Nearly Human: The Gorilla's Guide to Good Living* (Suffern, NY, 2007)
Haber, Karen, ed., *Kong Unbound: The Cultural Impact, Pop Myths, and Scientific Plausibility of a Cinematic Legend* (New York, 2005)
Haraway, Donna, *Primate Visions: Gender, Race, and Nature in the World of Modern Science* (New York and London, 1989)
Jones, Jeannette Eileen, '"Gorilla Trails in Paradise": Carl Akeley, Mary Bradley, and the American Search for the Missing Link', *The Journal of American Culture*, XXIX/3 (September 2006)
Nellemann, Christian, and Adrian Newton, eds, *Great Apes – The Road Ahead* (United Nations Environment Project, 2002), PDF available at www.unep-wcmc.org
—, Ian Redmond and Johannes Refisch, eds, *The Last Stand of the Gorilla: Environmental Crime and Conflict in the Congo Basin, A Rapid Response Assessment* (United Nations Environment Programme, GRID-Arendal, 2010), PDF available at www.unep.org
Patterson, Francine, and Wendy Gordon, 'The Case for the Personhood of Gorillas', in *The Great Ape Project*, ed. Paola Cavalieri and Peter Singer (New York, 1993)
Plowden, Gene, *Gargantua: Circus Star of the Century* (New York, 1972)
Robbins, Martha M., Pascale Sicotte and Kelly J. Stewart, eds, *Mountain Gorillas: Three Decades of Research at Karisoke* (Cambridge, 2001)
Savage, Thomas S., and Jeffries Wyman, 'Notice of the External Characters and Habits of *Troglodytes Gorilla*, a New Species of Orang from the Gaboon River, Osteology of the Same', *Boston Journal of Natural History*, V/4 (December 1847)
Schaller, George B., *The Mountain Gorilla: Ecology and Behaviour* [1963] (Chicago, 2010)
—, *The Year of the Gorilla* (Chicago, IL, 1964)
Schroeder, Barbet, *Koko: A Talking Gorilla* (Criterion DVD, 1978)

Associations and Websites

GORILLA CONSERVATION

Organizations working to protect gorillas include the following:

THE APE ALLIANCE
www.4apes.com/gorilla

BERGGORILLA & REGENWALD DIREKTHILFE E.V.
www.berggorilla.org

THE DIAN FOSSEY GORILLA FUND INTERNATIONAL
http://gorillafund.org

THE GORILLA ORGANIZATION
www.gorillas.org

GREAT APES SURVIVAL PARTNERSHIP
www.un-grasp.org

INTERNATIONAL GORILLA CONSERVATION PROGRAMME
www.igcp.org

VIRUNGA NATIONAL PARK, DEMOCRATIC REPUBLIC OF CONGO
http://gorillacd.org

Acknowledgements

For their friendship, research assistance and moral support of this project the authors would like to thank Michael Baxter, David Belzycki, Shane Carmody, Joan and Peter Clemenger, Isobel Crombie, Dario Gamboni, Luca Giuliani, Kevin and Maurene Gott, Robert Gott, Tom Gott, Doug Hall, Michael R. Leaman, Richard Linden, Frances Lindsay, Linda Mangan, eX de Medici, Jennie Moloney, Megan Patty, Robert M. Peck, Emil Giuliani Weir, James Robin-Weir and Peter Zegers.

For their help and advice with sourcing images for this publication, we would also like to thank the following individuals: Karl Amman, Diana Boston, Jean-Pierre Chabrol, Bob Deis, Ben Frost, Raymond Gill, H., C. & A. Glad, Greg Holfeld, Jacques de Loustal, Claudio Onorato, Ruth Starke, Marco Stepniak, Wade Williams, Doug Wilson and Aneurin Wright. As well as the following institutions and staff: American Museum of Natural History, New York, and Gregory August Rami; Art Africain-Masques d'Afrique, Furiani, France (Corsica); Australian War Memorial, Canberra, Lola Wilkins and Laura Webster; The Creature Technology Company, Melbourne, and John Barcham; The Dian Fossey Gorilla Fund International (www.gorillafund.org) and Barbara Joye; Library of Congress, Washington, DC, and Paul Hogroian; The Gorilla Foundation / Koko.org and Lorraine Slater; The Martin and Osa Johnson Safari Museum, Kansas, and Conrad Froehlich; National Gallery of Victoria, Melbourne, and Jennie Moloney; National Geographic and Susan Henry; National Library of Australia, Canberra, and Vicki Jovanovski; Norman Lindsay Estate and Barbara Mobbs; Pittsburg State University, Kansas, and Chris Kelly; Queensland Museum Library,

Brisbane, and Kathy Buckley; Queensland Art Gallery, Brisbane; Queensland Art Gallery Research Library, Brisbane, and Judy Gunning; Ringling Brothers and Barnum & Bailey Circus, Sarasota, FL, and Feld Entertainment Inc., and Heidi San Nicolas; Bibliothek and Bildarchiv, Secession, Vienna, and Astrid Steinbacher; State Library of Victoria, Melbourne, and Shane Carmody; The Travelers Companies, Hartford, CT, and Beverley A. Ripple; University of Queensland Library, Brisbane, and K. Kerswell; Victoria and Albert Museum, London, and Anna Sheppard.

Photo Acknowledgements

The authors and the publishers wish to express their thanks to the following sources of illustrative material and/or permission to reproduce it.

Photo courtesy of American Museum of Natural History, New York: p. 61; photograph courtesy of Karl Amman (http://karlammann.com): p. 24; courtesy of Art Africain – Masques d'Afrique (www.art-masque-africain.com): p. 22; Australian War Memorial, Canberra: pp. 167, 169 (courtesy of H., C. & A. Glad), 170; Kabir Bakie: p. 10 top; courtesy of Diana Boston: p. 119; Jean-pierre Chabrol: p. 156; courtesy of The Dian Fossey Gorilla Fund International: pp. 8 (KRC Veronica Vecellio), 11 (KRC Felix Ndagijimana), 52 (KRC Dean Jacobs), 73 (KRC Veronica Vecellio), 80; courtesy of Ben Frost: p. 143; Ted Gott: pp. 141, 154; iStockphoto: pp. 9 (Aleksandar Todorovic), 13 (Karen Givens), 15, 85 (Ronald van der Beek), 108 (MoMorad), 182 (David Morgan); courtesy of The Gorilla Foundation/Koko.org: p. 189 (Ron Cohn); courtesy of Greg Holfeld: p. 95 left; Julie Langford: p. 10 centre; Library of Congress, Washington, DC: p. 168; courtesy of Jacques de Loustal: p. 175; The Martin and Osa Johnson Safari Museum, Kansas: p. 67; collection of

MiM – Museum in Motion, San Pietro in Cerro, Piacenza, courtesy of Claudio Onorato (www.claudioonorato.it): p. 145; National Gallery of Victoria, Melbourne: p. 123; National Geographic Stock: p. 78 (Robert I. M. Campbell); National Library of Australia, Canberra: p. 88 left; Queensland Art Gallery Research Library, Brisbane: pp. 21, 164, 166; Queensland Museum Library, Brisbane: pp. 34, 159, 161; courtesy of Ringling Brothers and Barnum & Bailey Circus, Sarasota and Feld Entertainment, Inc.: pp. 115, 116; courtesy of Simon Schluter and *The Age*: p. 148; Shutterstock: p. 6 (Kamil Macniak); State Library of Victoria, Melbourne: pp. 26, 33, 35, 46, 88 right, 160, 162; courtesy of Marco Stepniak (www.stepniak-bild.de): p. 187; Secession Building, Vienna: p. 16; Travelers Archives, reproduced with permission: p. 151; Victoria and Albert Museum, London: p. 37; photographs for *Robot Monster*, 1953, *Bela Lugosi Meets a Brooklyn Gorilla*, 1952, and *Bride of the Gorilla*, 1951, licensed through the Wade Williams Collection: pp. 29, 30 top, 192; courtesy of Doug Wilson Wonderland Glass and Photography Studios: p. 181; courtesy of Aneurin Wright: p. 96.

Index